Advances in the Catalytic Conversion of Biomass Components to Ester Derivatives: Challenges and Opportunities

Advances in the Catalytic Conversion of Biomass Components to Ester Derivatives: Challenges and Opportunities

Editor

Domenico Licursi

MDPI • Basel • Beijing • Wuhan • Barcelona • Belgrade • Manchester • Tokyo • Cluj • Tianjin

Editor
Domenico Licursi
University of Pisa
Italy

Editorial Office
MDPI
St. Alban-Anlage 66
4052 Basel, Switzerland

This is a reprint of articles from the Special Issue published online in the open access journal *Catalysts* (ISSN 2073-4344) (available at: https://www.mdpi.com/journal/catalysts/special_issues/Catalytic_Conversion_Biomass).

For citation purposes, cite each article independently as indicated on the article page online and as indicated below:

LastName, A.A.; LastName, B.B.; LastName, C.C. Article Title. *Journal Name* **Year**, *Volume Number*, Page Range.

ISBN 978-3-0365-4121-1 (Hbk)
ISBN 978-3-0365-4122-8 (PDF)

© 2022 by the authors. Articles in this book are Open Access and distributed under the Creative Commons Attribution (CC BY) license, which allows users to download, copy and build upon published articles, as long as the author and publisher are properly credited, which ensures maximum dissemination and a wider impact of our publications.
The book as a whole is distributed by MDPI under the terms and conditions of the Creative Commons license CC BY-NC-ND.

Contents

About the Editor vii

Domenico Licursi
Advances in the Catalytic Conversion of Biomass Components to Ester Derivatives: Challenges and Opportunities
Reprinted from: *Catalysts* 2022, *12*, 455, doi:10.3390/catal12050455 1

Claudia Antonetti, Domenico Licursi and Anna Maria Raspolli Galletti
New Intensification Strategies for the Direct Conversion of Real Biomass into Platform and Fine Chemicals: What Are the Main Improvable Key Aspects?
Reprinted from: *Catalysts* 2020, *10*, 961, doi:10.3390/catal10090961 5

Alex de Nazaré de Oliveira, Erika Tallyta Leite Lima, Eloisa Helena de Aguiar Andrade, José Roberto Zamian, Geraldo Narciso da Rocha Filho, Carlos Emmerson Ferreira da Costa, Luíza Helena de Oliveira Pires, Rafael Luque and Luís Adriano Santos do Nascimento
Acetylation of Eugenol on Functionalized Mesoporous Aluminosilicates Synthesized from Amazonian *Flint* Kaolin
Reprinted from: *Catalysts* 2020, *10*, 478, doi:10.3390/catal10050478 21

Claudia Antonetti, Samuele Gori, Domenico Licursi, Gianluca Pasini, Stefano Frigo, Mar López, Juan Carlos Parajó and Anna Maria Raspolli Galletti
One-Pot Alcoholysis of the Lignocellulosic *Eucalyptus nitens* Biomass to *n*-Butyl Levulinate, a Valuable Additive for Diesel Motor Fuel
Reprinted from: *Catalysts* 2020, *10*, 509, doi:10.3390/catal10050509 49

Anna Maria Raspolli Galletti, Domenico Licursi, Serena Ciorba, Nicola Di Fidio, Valentina Coccia, Franco Cotana and Claudia Antonetti
Sustainable Exploitation of Residual *Cynara cardunculus* L. to Levulinic Acid and *n*-Butyl Levulinate
Reprinted from: *Catalysts* 2021, *11*, 1082, doi:10.3390/catal11091082 71

Ligang Luo, Xiao Han and Qin Zeng
Hydrogenative Cyclization of Levulinic Acid to γ-Valerolactone with Methanol and Ni-Fe Bimetallic Catalysts
Reprinted from: *Catalysts* 2020, *10*, 1096, doi:10.3390/catal10091096 91

Zhiyi Wu, Pingzhou Wang, Jie Wang and Tianwei Tan
Guerbet Reactions for Biofuel Production from ABE Fermentation Using Bifunctional Ni-MgO-Al_2O_3 Catalysts
Reprinted from: *Catalysts* 2021, *11*, 414, doi:10.3390/catal11040414 105

Luigi di Bitonto, Valeria D'Ambrosio and Carlo Pastore
A Novel and Efficient Method for the Synthesis of Methyl (*R*)-10-Hydroxystearate and FAMEs from Sewage Scum
Reprinted from: *Catalysts* 2021, *11*, 663, doi:10.3390/catal11060663 119

Valdis Kampars, Ruta Kampare and Aija Krumina
MgO Catalysts for FAME Synthesis Prepared Using PEG Surfactant during Precipitation and Calcination
Reprinted from: *Catalysts* 2022, *12*, 226, doi:10.3390/catal12020226 135

About the Editor

Domenico Licursi obtained his master's degree in Industrial Chemistry in 2010 and PhD in Chemical Sciences from Galileo Galilei Graduate School in 2015. He is working as a Researcher in Industrial Chemistry at the Department of Chemistry and Industrial Chemistry, University of Pisa. He has investigated acid-catalyzed hydrothermal biomass conversion into added-value bioproducts, mainly 5-hydroxymethylfurfural and levulinic acid, as well as the catalytic production of some of their key oxygenated derivatives (alkyl levulinates, ketals, gamma-valerolactone, 2-methyltetrahydrofuran, furan diols). His current research concerns the hydrothermal carbonization and/or pyrolysis of waste biomasses to produce hydro-/biochars and bio-oils that are potentially exploitable for fuel and environmental applications. He is an active member of the American Chemical Society (ACS), mainly within the Agrochemicals, Catalysis Science & Technology, and Cellulose & Renewable Materials divisions.

Editorial

Advances in the Catalytic Conversion of Biomass Components to Ester Derivatives: Challenges and Opportunities

Domenico Licursi

Department of Chemistry and Industrial Chemistry, University of Pisa, Via Giuseppe Moruzzi 13, 56124 Pisa, Italy; domenico.licursi@unipi.it

Citation: Licursi, D. Advances in the Catalytic Conversion of Biomass Components to Ester Derivatives: Challenges and Opportunities. *Catalysts* 2022, 12, 455. https://doi.org/10.3390/catal12050455

Received: 12 April 2022
Accepted: 14 April 2022
Published: 20 April 2022

Publisher's Note: MDPI stays neutral with regard to jurisdictional claims in published maps and institutional affiliations.

Copyright: © 2022 by the author. Licensee MDPI, Basel, Switzerland. This article is an open access article distributed under the terms and conditions of the Creative Commons Attribution (CC BY) license (https://creativecommons.org/licenses/by/4.0/).

Sustainable conversion of biomass feedstocks into valuable bio-fuels and bio-products plays a strategic role within modern industrial catalysis. However, despite the diversified possibilities already available and well developed on a laboratory scale, it is necessary to focus the potentially interesting bio-products at the next industrial scale, in the perspective of the concrete realization of the coming process intensification. The final application of a bio-product represents an essential prerequisite to define its potential industrial interest, but this is not enough to justify its effective production on a larger scale. To solve this bottleneck, one-pot biomass conversion approaches, carried out under the mildest conditions, using robust catalysts, and high biomass loadings, should be preferred. In this context, the esterification pathway fully fits all these requirements, and the exploitation of ester derivatives is already attractive in many fields of applied and industrial research. These aspects are of paramount importance and have been discussed more in-depth in the introductory Editorial of this Special Issue, focused on the identification of the main improvable key aspects for achieving the intensification of the esterification processes, but extendable to many other biomass conversion strategies [1].

In the first work of this Special Issue, de Oliveira et al. [2] have investigated the acetylation of eugenol with acetic anhydride, deepening the catalytic activity of a mesoporous catalyst, synthesized from 3-mercaptopropyltrimethoxysilane functionalized Amazonian *flint* kaolin, a kaolinitic waste residue disposed in the mine shortly after exploration. This work is well-supported by the value of the final product, eugenyl acetate, which can be used as a natural and efficient larvicide. In addition, the authors rightly claim some typical strengths, such as good stability, recyclability, and high catalytic activity of their catalyst, which was synthesized following the criteria of simplicity, low cost, and environmental sustainability. As a concluding remark, I highlight the well-developed discussion about the characterization of the synthesized catalyst.

Another valuable exploitation possibility of ester derivatives is certainly in the field of the oxygenated bio-fuels, and alkyl levulinates are attractive for this purpose. Antonetti et al. [3] have studied the one-pot butanolysis of the autohydrolyzed-delignified *Eucalyptus nitens* wood to *n*-butyl levulinate, adopting *n*-butanol as the green reagent/reaction medium, very dilute sulfuric acid as the homogeneous catalyst, and comparing microwave and traditional heating systems. To achieve the optimization of this reaction, the authors have developed a face-centered central composite design, experimentally validating their model at the optimal operating conditions for *n*-butyl levulinate production. In addition, a preliminary study of diesel engine performances and emissions for a model mixture with a composition analogous to that of the main components of the reaction mixture, has been proposed, to draw an indication of its potential application as an additive for diesel fuel, resulting in a relevant reduction in CO and soot emission. Certainly, the main strength of this work is the well-balanced synergy between catalysis and the development of the butyl levulinate application as a novel oxygenated bio-fuel, which is a topic of great interest and actuality. Moreover, the high biomass loading (20 wt%) adopted in the butanolysis reaction represents an eligible aspect from the perspective of the high gravity approach, which is conducive to the next process intensification. Similar conclusions have been claimed by Raspolli et al. [4] for the conversion to *n*-butyl levulinate of a *Cynara cardunculus* L. (cardoon) waste residue, obtained after seed removal for oil exploitation. Additionally, MW-assisted hydrolysis of the same biomass to levulinic acid was also investigated and optimized. In this context, an interesting section of the manuscript deals with the characterization of reaction by-products and the purification of the synthesized levulinic acid. The achieved results confirm that eco-friendly reaction conditions can be adopted for the conversion of low-cost residual cardoon both to the carboxylic acid or to its ester derivative.

Even cyclic esters, such as γ-valerolactone, are receiving considerable attention as bio-fuels, food additives, solvents, and drug intermediates. In the specific case of γ-valerolactone, its synthesis occurs via reduction and lactonization of the above mentioned levulinic acid, but it requires the development of low-cost, performing, and easily regenerable catalysts. Developing this approach, Luo et al. [5] have synthesized a series of Ni-Fe/SBA-15 catalysts, testing them for the catalytic hydrogenation of levulinic acid to γ-valerolactone, carried out in the presence of methanol as the only hydrogen donor, and investigating the synergism between Fe and Ni towards this reaction. The good catalytic performances/stability, the low cost, and the easy synthesis/regeneration of the synthesized Ni-Fe bimetallic catalysts are key aspects for the real development of γ-valerolactone production on a larger scale, thus contributing to fill the existing gap between the academic and industrial world.

From a different perspective, Wu et al. [6] have synthesized and characterized novel MgO–Al_2O_3 mixed metal oxides decorated with Ni nanoparticles (Ni–MgO–Al_2O_3), which have been tested on the Guerbet condensation reaction of ABE mixture into long-chain (C5–C15) ketones and alcohols, which are important bio-fuel precursors. Ni nanoparticles facilitate the dehydrogenation/hydrogenation process, whilst the activity of the condensation reactions is highly related to the acidity and basicity of the catalyst, which requires a careful balance and optimization, depending on the desired product(s) of interest. In their work, the authors have optimized the production of C5–C15 ketones and alcohols, after optimization of the Mg/Al ratio and Ni loading. However, the wide modularity of the involved ternary catalytic systems could be advantageously exploited to select the Guerbet reaction towards the selective production of aldo-esterification products.

Bio-diesel production certainly represents another very hot topic in the field of esterification reactions. It is a biodegradable and renewable fuel, showing chemical and physical properties similar to those of the petroleum-based fuels, and includes a mixture of fatty acid methyl esters (FAMEs), but also other main other components, hitherto little exploited. Applying the fractionation concept on the lipid component, Di Bitonto et al. [7] have proposed the one-pot transesterification of methyl estolides extracted from the lipid component of the sewage scum with methanol, to give methyl 10-(R)-hydroxystearate and FAMEs. Transesterification has been optimized by the authors developing a three-level and four factorial Box–Behnken experimental design, using $AlCl_3 \cdot 6H_2O$ or HCl as the catalyst. In both cases, a complete conversion of methyl estolides into methyl 10-(R)-hydroxystearate and FAMEs, was ascertained. The products have been isolated, quantified, and fully characterized. At the end of the process, methyl 10-(R)-hydroxystearate was purified and well-characterized by NMR spectroscopy. The high enantiomeric excess (>92%) of the isolated R-enantiomer isomer opens a new scenario for the valorization of sewage scum towards a multi-product sustainable biorefinery, thus achieving the complete valorization of the lipid fraction present in the starting feedstock. In fact, in addition to FAMEs, already well known as green and performing bio-fuels, methyl 10-(R)-hydroxystearate can be a valuable source of 10-(R)-hydroxystearic acid, a chemical used in the manufacturing of lubricants and cosmetics, and this new approach smartly solves the drawbacks of its current synthesis by enzymatic hydrolysis of edible vegetable oils.

In the last contribution, Kampars et al. [8] have proposed the use of MgO as the catalyst for bio-diesel synthesis. MgO synthesis was carried out starting from magnesium nitrate with ammonia, and calcination was performed at different temperatures in air, in the presence of PEG as surfactant and fuel, leading to active catalysts towards this reaction. For most of these catalysts, FAME yield increased as the size of the crystallites of the catalyst decreased. FTIR spectra showed that deviations in this relationship may be due to the retention of incomplete calcination products on the surface of MgO. FAME and intermediate yield dependence on oil conversion confirmed that all catalysts had the same types of strong base sites that were necessary for the initialization of the transesterification reactions. In my opinion, the proposal of MgO as a catalyst for bio-diesel production well fits the general requirements and trends of the industrial catalysis, being eligible for activity, stability, wide availability, low cost, and environmental sustainability.

In summary, these eight papers highlight the relevance of obtaining ester derivatives for developing applications of great industrial interest, fully according to sustainability and cost-effectiveness criteria. I would like to thank all the authors of this Special Issue for their contributions, as well as all the reviewers for having improved the quality of the submitted papers with their valuable comments. I am also grateful to all the staff of the Catalysts Editorial Office for their helpful suggestions.

Funding: This research received no external funding.

Conflicts of Interest: The author declares no conflict of interest.

References

1. Antonetti, C.; Licursi, D.; Raspolli Galletti, A.M. New intensification strategies for the direct conversion of real biomass into platform and fine chemicals: What are the main improvable key aspects? *Catalysts* **2020**, *10*, 961. [CrossRef]
2. De Nazaré de Oliveira, A.; Lima, E.T.L.; de Aguiar Andrade, E.H.; Zamian, J.R.; da Rocha Filho, G.N.; da Costa, C.E.F.; de Oliveira Pires, L.H.; Luque, R.; do Nascimento, L.A.S. Acetylation of eugenol on functionalized mesoporous aluminosilicates synthesized from Amazonian flint kaolin. *Catalysts* **2020**, *10*, 478. [CrossRef]
3. Antonetti, C.; Gori, S.; Licursi, D.; Pasini, G.; Frigo, S.; Lopez, M.; Parajo, J.C.; Raspolli Galletti, A.M. One-pot alcoholysis of the lignocellulosic *Eucalyptus nitens* biomass to *n*-butyl levulinate, a valuable additive for diesel motor fuel. *Catalysts* **2020**, *10*, 509. [CrossRef]
4. Raspolli Galletti, A.M.; Licursi, D.; Ciorba, S.; di Fidio, N.; Coccia, V.; Cotana, F.; Antonetti, C. Sustainable exploitation of residual *Cynara cardunculus* L. to levulinic acid and *n*-butyl levulinate. *Catalysts* **2021**, *11*, 1082. [CrossRef]
5. Luo, L.; Han, X.; Zeng, Q. Hydrogenative cyclization of levulinic acid to γ-valerolactone with methanol and Ni-Fe bimetallic catalysts. *Catalysts* **2020**, *10*, 1096. [CrossRef]
6. Wu, Z.; Wang, P.; Wang, J.; Tan, T. Guerbet reactions for biofuel production from ABE fermentation using bifunctional Ni-MgO-Al_2O_3 catalysts. *Catalysts* **2021**, *11*, 414. [CrossRef]
7. Di Bitonto, L.; D'Ambrosio, V.; Pastore, C. A novel and efficient method for the synthesis of methyl (R)-10-hydroxystearate and FAMEs from sewage scum. *Catalysts* **2021**, *11*, 663. [CrossRef]
8. Kampars, V.; Kampare, R.; Krumina, A. MgO catalysts for FAME synthesis prepared using PEG surfactant during precipitation and calcination. *Catalysts* **2022**, *12*, 226. [CrossRef]

Editorial

New Intensification Strategies for the Direct Conversion of Real Biomass into Platform and Fine Chemicals: What Are the Main Improvable Key Aspects?

Claudia Antonetti, Domenico Licursi * and Anna Maria Raspolli Galletti

Department of Chemistry and Industrial Chemistry, University of Pisa, Via Giuseppe Moruzzi 13, 56124 Pisa, Italy; claudia.antonetti@unipi.it (C.A.); anna.maria.raspolli.galletti@unipi.it (A.M.R.G.)
* Correspondence: domenico.licursi@unipi.it; Tel.: +39-050-221-0543

Received: 7 August 2020; Accepted: 20 August 2020; Published: 22 August 2020

Abstract: Nowadays, the solvothermal conversion of biomass has reached a good level of development, and now it is necessary to improve the process intensification, in order to boost its further growth on the industrial scale. Otherwise, most of these processes would be limited to the pilot scale or, even worse, to exclusive academic investigations, intended as isolated applications for the development of new catalysts. For this purpose, it is necessary to improve the work-up technologies, combining, where possible, reaction/purification unit operations, and enhancing the feedstock/liquid ratio, thus improving the final concentration of the target product and reducing the work-up costs. Furthermore, it becomes decisive to reconsider more critically the choice of biomass, solvent(s), and catalysts, pursuing the biomass fractionation in its components and promoting one-pot cascade conversion routes. Screening and process optimization activities on a laboratory scale must be fast and functional to the flexibility of these processes, exploiting efficient reaction systems such as microwaves and/or ultrasounds, and using multivariate analysis for an integrated evaluation of the data. These upstream choices, which are mainly of the chemist's responsibility, are fundamental and deeply interconnected with downstream engineering, economic, and legislative aspects, which are decisive for the real development of the process. In this Editorial, all these key issues will be discussed, in particular those aimed at the intensification of solvothermal processes, taking into account some real case studies, already developed on the industrial scale.

Keywords: process intensification; alcoholysis; hydrolysis; solvothermal process; alkyl levulinate; levulinic acid; 5-hydroxymethylfurfural; furfural; humins

---

1. Importance of the Solvothermal Processes for the Synthesis of Platform and Fine Chemicals

The use of biomass for the production of platform and fine chemicals is strongly imposing as a valid alternative to the employment of traditional fossil sources [1]. However, despite the progress achieved in the optimization and development of many carbohydrate-based processes, it is still necessary to invest resources in research and development, to overcome the significant gap between the academic and the industrial world, and this is possible only by improving the process intensification [2]. Process intensification favors equipment size reductions, leading to enhancement in chemical reaction kinetics, energy efficiency, process safety, minimization of waste generation, and reduction in capital costs. Maximum atom-economy and minimum energy requirements are fundamental features for the development of more sustainable and greener processes. For example, in the hydrothermal processes, which fall into the broader solvothermal processes, the use of sub- and supercritical water, as the reaction medium, provides a valuable and sustainable path for reducing the use of organic solvents. This technology can be integrated for the development of safer, more flexible, economical, and ecological biomass conversion processes, in particular carbonization, liquefaction, and gasification, which differ mainly in the adopted pressure, temperature, and residence time [3]. Moreover, hydrothermal technology offers the advantage of directly converting raw biomasses, also those with relatively high-moisture content, without any energy-intensive pretreatment [4]. In the last years, increasing research has been devoted to the optimization of hydrothermal processes, in particular under sub-critical conditions, in the absence or presence of a suitable acid catalyst, for the selective production of reducing sugars or platform chemicals. Some noteworthy examples are the furanic intermediates, such as furfural and 5-hydroxmethylfurfural, deriving from the dehydration of C5 and C6 carbohydrates, respectively [5], and organic acids (levulinic plus formic acids), resulting from the rehydration of C6 furanic intermediates [6]. Most recent advances in the production of these valuable platform chemicals will be discussed in the following paragraph.

2. Some Relevant Examples of C5 and C6 Derivatives of Industrial Interest

Nowadays, furfural is exclusively produced from the acid-catalyzed hydrolysis of the pentosan fraction of lignocellulosic biomasses to xylose and its subsequent dehydration. Its industrial process provides one or two separate steps, carried out in batch or continuous reactors, and in the presence of a mineral acid as the reaction catalyst [7]. Due to the quick degradation of furfural occurring in the liquid phase, it must be readily removed once synthesized and, for this purpose, traditional separation techniques can be adopted, such as steam or nitrogen stripping, supercritical carbon dioxide extraction, mono/biphasic solvent extraction, and adsorption on resins. Moreover, on an industrial scale, significant technological improvements have been achieved, always aimed at minimizing its residence time in the liquid phase, enhancing its concentration in vapor products, and effectively reusing the acid catalyst [7]. Going towards a more intensified process development, also reactive distillation has been proposed, adopting zeolite H-mordenite as the solid catalyst and xylose as the starting feedstock [8]. In this way, the immediate separation of furfural from the reaction system can be advantageously achieved, minimizing the formation of undesired condensation products and leading to significant heat integration benefits. The furfural market size is projected to grow from EUR 500 million in 2019 to EUR 630 million by 2024, with Austria, Belgium, China, Dominican Republic, India, Italy, Slovenia, South Africa, and US being the key market players, and with furfuryl alcohol, tetrahydrofuran, maleic anhydride, 2-methyl tetrahydrofuran, and 1,5-pentanediol as the main derivatives, already manufactured on a commercial scale [5]. Further, in the case of 5-hydroxymethylfurfural (5-HMF) production, the formation of by-products, separation, and purification issues, as well as catalyst regeneration, have been identified as major challenges [9]. To solve these drawbacks, many efficient solutions have been continuously proposed, such as that of Yan et al. [10], who have performed the continuous production of 5-HMF in a flow-reactor, achieving high yields, starting from fructose or glucose, using HCl and $AlCl_3$ as the catalysts and water as the reaction medium. Moreover, in this case, high yields can be reached thanks to a very efficient extraction step, which allows the continuous removal of this reactive furan from the reaction environment: the produced 5-HMF is extracted into the organic phase in real time, thus avoiding its further degradation, while the unreacted carbohydrates re-enter spontaneously into the reaction phase for another reaction cycle. Moreover, the authors have performed a techno-economic analysis, demonstrating that 5-HMF could be produced at a minimum selling price of USD 1716/ton and USD 1215/ton from fructose and glucose, respectively, which brings motivation and a real chance for its further commercial production in greater volumes. The industrial production of 5-HMF is already available and carried out by AVA Biochem [11], with a monophasic water-based hydrothermal process, which has been optimized and acknowledged as sustainable, efficient, and robust and certainly economically advantageous. In recent studies of environmental sustainability assessment, the concrete feasibility of the 5-HMF oxidation to 2,5-furandicarboxylic acid has been also demonstrated [12,13]. The latter is one of the 12 uppermost chemical building blocks, which can be used for the production of polymers and resins, such as polyethylene furanoate, which is a promising substitute for polyethylene-terephthalate [14]. Acid-catalyzed hydrolysis of 5-HMF leads to levulinic acid, another valuable platform chemical of great interest, and also an assessment of this hydrothermal process has been recently evaluated and discussed, demonstrating its environmentally friendly and neutral safety performances [15]. The interest in the scale-up of these hydrothermal processes on an industrial scale is further strengthened by the continuous development of new conversion strategies of the above platform chemicals into more added-value fine chemicals, which are industrially more attractive, due to their ready-to-use applications [16,17]. However, considerable improvement opportunities are possible, in particular finding more environmentally friendly solvents for performing the reactions and the recovery of the desired product(s) and developing new catalysts, aimed at improving the process efficiency and reducing the energy consumption. Besides the water medium, involved in the hydrothermal approach, solvothermal processes can be also performed with other green and sustainable solvents, such as alcohols, in some cases attaining remarkable advantages. Taking into account the hydrothermal process for levulinic acid production as the reference example, the main advantages of the alcoholysis route consist of (i) the development of more value-added products, the alkyl levulinates, now exploitable as oxygenated additives for gasoline and diesel fuels; (ii) an easier work-up procedures, generally by distillation, thanks to the lower boiling points of the esters; and (iii) a reduced number of process units and enhanced performances of new technological solutions, such as reactive distillation, conducive to the process intensification [18]. In this way, it is possible to combine reaction and separation unit operations, allowing simpler, more efficient, economical, and cleaner production processes. Further, in this case, worldwide techno-economic and environmental assessment of alkyl levulinates production has been recently proposed, highlighting a promising economic outlook of these bio-products [19].

Although solvothermal biomass processing has been recognized as a really promising technology for converting lignocellulosic and waste biomasses into valuable bio-chemicals and bio-fuels, some key aspects should be still considered, for improving the development of their industrial applications, in particular in the perspective of the intensification process development. Some common drawbacks of these processes are already well known, in particular the corrosion of the equipment, due to the use of concentrated mineral acids for the catalysis, the precipitation of inorganic salts of biomass source in the presence of sulfuric acid as acid catalyst, char/coke formation, associated with the presence of unconverted biomass, and thermal decomposition of the bio-oil [20]. However, most of these problems can be solved after an appropriate optimization study, preferring the use of heterogeneous catalysts in the presence of a soluble substrate (for instance, when beverage or sugar industry waste is converted) and, if this is not possible, adopting very dilute mineral acids, which must be properly recovered. In the next paragraphs, we discuss more in detail about some additional choices, which should be carefully done upstream of the process development, to significantly improve the next intensification phase for the production of C5 and C6 derivatives on a larger scale.

3. Choose Strategic Reaction Components

3.1. About the Starting Biomass

First of all, the efficiency of the hydrothermal-solvothermal processes can be improved upstream, choosing a starting biomass feedstock with a suitable chemical composition, depending on the chemical process to develop. Given the importance of the catalytic upgrading of C5 and C6 sugars, a promising biomass feedstock for the production of biofuels and bioproducts should have a good content of carbohydrates and low recalcitrance to their conversion, which is favored by a low lignin content [21–23]. Besides, biomasses of low cost, low input cultivation, and wide availability in the territory should be preferred, even in laboratory investigations [24]. Lignocellulosic biomass comprising of agricultural and forest residues (such as wheat straw, rice straw, rice husk, corn stover, sugarcane bagasse) and energy crops are extremely attractive for these purposes. Even more so, at an advanced stage of process intensification, an economic analysis of biomass supply chains, including collection, processing, and transport, is necessary for identifying the best plant locations that balance economic, environmental, and social criteria, making all actors (farmers, investors, industrial entrepreneurs, government) aware that success relies on agreement advances [25,26]. Flexible hydrothermal processes are advisable, allowing the use of different kinds of biomass, an aspect of paramount importance taking into account their certain seasonality. Even waste biomasses, such as food and cooked food waste, can be effectively exploited to platform chemicals, compensating for seasonality issues of the lignocellulosic biomasses [27]. Cellulosic waste materials from papermaking processes, which cannot be re-used for the production of new paper, are valuable feedstocks for the synthesis of bio-products, such as levulinic acid, thus saving the costs deriving from their traditional disposal in landfills or from their use in waste-to-energy plants [28]. Hydrothermal treatment of sewage sludge is a promising strategy for sustainable management, allowing its conversion into useful products, and simultaneously mitigating the environmental risks [29]. In this case, besides the catalyst recovery issue, the recirculation of the liquid effluent within the process could overcome the legal thresholds, such as chemical oxygen demand (COD) and heavy metals [30,31]. These possible drawbacks suggest that European and/or national legislation improvements are still necessary for allowing the development of this process intensification [30,31].

3.2. About the Reaction Medium

Hydrothermal technology involves the use of water as the preferred green reaction solvent, promoting the process sustainability, and shifting the attention towards the development of more efficient and economical work-up strategies, such as solvent extraction, distillation, and/or membrane separation. Process intensification provides the combination of multiple process tasks or equipment into a single unit and the development of material/energy integration, as occurs for the production of furans, such as furfural [32] and 5-HMF [9], and organic acids [33], such as the previously mentioned levulinic acid [34,35]. Subcritical water represents a promising reaction medium for successful biomass exploitation, due to its interesting physicochemical properties, at the typical reaction conditions [36]. However, the replacement of water with an alcoholic solvent should lead to bio-products of higher added-value, such as alkyl-glucosides/xylosides [37,38], alkoxymethyl furfural [39], or the previously mentioned alkyl levulinates [40]. To improve the poor solubility of biomass in water and organic solvents, ionic liquids and deep eutectic solvents can be effectively used, exploiting their high solvation capacity towards the dissolution of carbohydrates and lignin, even in the case of the raw biomass [41,42]. However, ionic liquids still suffer from several disadvantages, having environmentally unfriendly aspects and requiring cost-intensive preparation procedures, whilst deep eutectic solvents solve many of these disadvantages, in many cases being green, environmental-friendly, and highly tunable. In particular, deep eutectic solvents have been recently proposed not only for the pretreatment of biomass, mainly aimed at the delignification and solubilization of cellulose and a decrease in its crystallinity degree, but also for their further conversion to added-value bio-chemicals [43]. Some noteworthy examples are cellulose modification by acetylation [44], cationic [45], and anionic functionalization [46], the dehydration of C5 and C6 carbohydrates to furfural and 5-HMF, respectively [47], and cellulose oxidation to gluconic acid [48]. However, despite the exciting performances, further research in this field is necessary, in particular for improving the separation and the purification of these deep eutectic solvents, further lowering their cost, and justifying their use on a larger scale.

3.3. About the Catalyst

Regarding the appropriate choice of catalyst, it should be made after that of the biomass, and properly tuned, based on the reaction of interest. For example, the acidity of high-pressure CO_2 can be advantageously exploited for biomass pre-treatment purposes, for improving the biomass digestibility before enzymatic hydrolysis, or performing the hemicellulose fractionation by mild autohydrolysis [49]. In both cases, cellulose and lignin components remain almost unaltered in the solid phase, which could be further fractionated and upgraded [49]. The use of CO_2 as an acid catalyst for biomass pre-treatments is certainly attractive, thanks to its non-toxicity, low cost, and ready availability, but its acidity is generally insufficient to perform harsher acid conversion routes, which require the use of stronger acid catalysts. From a practical and environmental perspective, it is imperative to develop heterogeneous catalysts that are hydrothermally stable at the process conditions. In this context, much work has been done, testing synthesized metal-based catalysts on many transformations of C5 and C6 carbohydrates involving the breakage of the C-O bonds for the synthesis of biofuels and bio-products, such as isomerization, dehydration, aldol condensation, ketonization, and hydrogenation [50,51]. Based on our experience, Nb-based catalysts are particularly promising for many of these purposes, showing excellent water tolerance, low cost, easy synthesis, tunable compositions, good acid properties (acid types, amount, and strength), and promising surface properties (specific surface areas, pore size, and volume) [52]. However, despite the numerous advances in the development of efficient heterogeneous catalysts, these often suffer from uninspiring performances due to mass transfer issues, deactivation due to cocking and water corrosion at the reaction conditions, clogging of the active pores, poisoning, and recycling issues. Consequently, even if new efficient catalysts are hourly synthesized and deeply characterized, their use for the synthesis of base chemicals has been mostly limited to academic investigations. On the other hand, the use of commercial catalysts, which are widely available and cheap, should be preferred for faster development of the bio-product applications, focusing rather on the improvement of other parameters related to the intensification process. The use of homogenous catalysts generally greatly improves biomass accessibility, giving back a better product yield/selectivity. For acid-catalyzed biomass conversion processes, the reactivity of a mineral catalyst is related to several prominent factors, such as its strength and concentration, type/loading of the biomass, and reaction conditions, in particular temperature and reaction time [36], and all these parameters should be properly considered and tuned, preferably by multivariate optimization. In the case of the levulinic acid production, the safety assessment due to the use of mineral acids does not present significant risks [15], whilst the catalyst recovery, together with the isolation/purification of the product(s), both with minimum energy input, still represent a challenging topic [6,20]. For example, the recovery of the acid catalyst within the levulinic acid process can be carried out by flash separation, if volatile hydrochloric acid is the chosen catalyst, whilst organic solvent extraction is still preferred to separate the product from the high-boiling sulfuric acid, which remains in the water solution, ready to be reprocessed [53]. Instead, regarding product isolation/purification, atmospheric/vacuum distillation and steam stripping are adopted for the separation of LA, which can be obtained with a final purity of about 95–97% [17]. However, the high boiling point of levulinic acid is not energetically favorable for distillation, and solvent extraction could be a viable alternative, but the high amounts of solvent which need to be evaporated make also this operation energy-intensive and costly [54]. Significant improvements are still possible in the work-up procedures, which are decisive for further lowering the production costs of these bio-products.

4. Fractionate and Exploit Each Biomass Component

4.1. Selective Biomass Fractionation

Another aspect to be improved for achieving the best process intensification is the selective fractionation of the biomass in its components, by optimizing each step, in agreement with the biorefinery concept [55]. For example, reactive furanic compounds, such as furfural, must be produced and recovered upstream of integrated processes, through very mild pre-treatments, and possibly applying the key concepts of the process intensification. In this context, Zang et al. [56] have proposed the biorefinery of switchgrass biomass for the integrated production of furfural, lignin, and ethanol. The chosen reaction system is composed of a biphasic solvent [choline chloride/methyl isobutyl ketone], with a deep eutectic solvent which enables the fast hemicellulose solubilization/conversion to furfural, which is simultaneously extracted by methyl isobutyl ketone. Regarding the fate of the remaining fractions, cellulose is subsequently converted into ethanol by enzymatic hydrolysis, whilst lignin is properly precipitated and recovered, for further added-value applications. Despite the use of a biphasic system and the improvable extraction/purification of the lignin fraction, the developed techno-economic analysis clarifies that the proposed biorefinery is still cost-competitive and has a low-economic risk, with the reaction temperature and the solid loading having the largest impacts on the minimum furfural selling price (estimated at 625 USD/t, in the best case of study). Another interesting approach has been proposed by Rivas et al. [57], who studied the hydrothermal conversion of eucalyptus (*Eucalyptus globulus*) to levulinic acid, the latter being further upgraded to γ-valerolactone. In the perspective of developing an integrated approach, a mild hydrothermal pretreatment was optimized upstream, which allowed the almost quantitative solubilization of thermolabile extractives and hemicelluloses, recovering a cellulose-rich solid, which was subjected to harsher hydrolysis to the desired levulinic acid.

4.2. Recovery/Exploitation of By-Products: New Trends

About the lignin component, many biomass-integrated biorefineries underestimate its importance, considering it as a not particularly valuable by-product, more similar to waste, rather than a resource. In the hydrothermal process for levulinic acid production, lignin is recovered as the main waste stream at the end of the process, being a carbonaceous solid residue, or hydrochar, whilst simple phenols are solubilized in the liquid phase. Regarding the solid hydrochar, it can be immediately used for energy recovery, which is generally impactful for these processes. Alternatively, this hydrochar, which is rich in hydroxylic and carboxylic functionalities [58,59], can be advantageously used to replace traditional fossil-based polyols, for example, for the formulation of flexible polyurethane foams [60], but also for applications as adsorbents, precursors of catalysts, soil amendment, anaerobic digestion and composting, and energy storage materials [61]. On the other hand, the solubilized phenols can be advantageously used as efficient antioxidants, as already demonstrated in the case of the hydrothermal treatment of the *Arundo donax* L. [62], and their separation from the other compounds can be well integrated with the available process technologies. The new lignin exploitation strategies are generally carried out on the lignin solid residue recovered from the hydrothermal process of levulinic acid, significantly improving the overall process economy, which is already rewarded by the levulinic acid production, at the same time minimizing the waste disposal.

On the other hand, the lignin recovery could be even more advantageous if carried out upstream of the integrated biorefinery, as provided by the available organosolv pretreatments, where an organic solvent is used to extract lignin in its native form [63]. High-purity cellulose selectively remains in the solid phase, but it is more prone to the following step of enzymatic/chemical hydrolysis, due to the increased contribution of its amorphous phase. Instead, the extracted lignin can be precipitated from the liquid phase by water dilution and recovered as a solid, while the hemicellulose fraction remains in the liquid stream, thus achieving an efficient fractionation of the biomass in its components. Focusing on the lignin fate, this component is less degraded, if compared with its downstream recovery, and therefore the "upstream" lignin is more useful for the development of higher-value applications, such as the production of carbon materials, vanillin and other oxidized compounds, phenolic antioxidants, bio-oil, BTX hydrocarbons, urethanes, epoxy resins, fire retardants, sequestering agents, nanomaterials, energy storage device, and many more [64]. The choice of "upstream" or "downstream" recovery of the different fractions depends on the value of the primary target product, which should pay off the whole process, and the possibility of obtaining economic surplus from secondary streams must be down-to-earth and supported by feasibility studies.

Although hydrothermal/solvothermal reactions can be improved and optimized by choosing appropriate reaction conditions and catalysts, some reaction by-products are inevitably formed. An example of great interest is given by the furanic humins, which are condensation products of C5 and C6 sources [65]. Their formation is particularly favored in the aqueous acid environment and under harsh reaction conditions, like those which typically occur for levulinic acid production, but also for that of reactive furanic compounds from simple C5 and C6 sugars, e.g., furfural and 5-HMF, respectively [66]. Instead, alcoholic solvents generally stabilize these soluble furans in the liquid phase, minimizing the next humin formation, growth, and precipitation [67], and therefore alcoholysis results more advantageous for improving the selectivity to the target product and the final carbon balance. As a further complication, if real lignocellulosic biomass is chosen as the starting feedstock for developing the conversion of interest, the resulting final solid hydrochar should include both the degraded lignin (as "pseudo-lignin") of the biomass source and humins, the latter being less important in the milder HTC processes [68]. All these by-products must be considered as a resource rather than waste of the process, and new exploitation strategies must be developed, to lower further the minimum selling price of the main bio-product(s) of interest, smartly completing the biomass biorefinery [69]. Up to now, the best-known applications of carbonaceous hydrochar/humins include energy production [70] and environmental remediation [71]. Besides, new possibilities have been recently proposed for the exploitation of humins, such as the synthesis of new biomaterials [72–74], syngas [75], and carboxylic acids [76,77]. Most of these applications have been developed by the Avantium Company [78], which produces 5-HMF on a pilot scale and a new class of furanic building blocks, called YXY, to use as bioplastics and biofuels, starting from first- and second-generation feedstocks [79]. Therefore, it is clear that the efficient and diversified exploitation of humins should improve the overall economy of their process. In our opinion, particular attention should be given to the catalytic conversion of humins by hydrotreatment, aimed at their liquefaction, depolymerization, and conversion into more valuable liquid chemicals, such as furanics, aromatics, and phenolics. For this purpose, Wang et al. [80] worked at 400 °C, using Ru/C as the hydrotreatment catalyst, and formic acid in isopropanol as the hydrogen donor, aimed at the selective production of substituted alkyl phenolics and higher oligomers, together with naphthalenes, and cyclic alkanes. Now, Sun et al. [81] have synthesized a Ru/W-P-Si-O bifunctional catalyst, testing it for the hydrotreatment of humins to give cyclic and aromatic hydrocarbons. The authors declare a high yield to cyclic hydrocarbons (up to 88.3%), working at 340–380 °C, and exploiting the cooperative catalysis between the nano-Ru particles and the strong Lewis acidity of the solid W-P-Si-O, the latter catalyzing the Diels–Alder reaction on the furan rings.

In the context of the developable applications of hydrochar/humins, the synthesis of furanic and carboxylic acid derivatives with catalysts directly developed from waste and by-products of the same hydrothermal/solvothermal processes represents a very hot topic, which should allow the improvement of the process intensification, also applying the biorefinery concept and zero waste policies. Besides, the heterogenization of the catalyst should be certainly advantageous for these processes, if the synthesized catalysts result as performing, their precursors have a low cost or, even better, a negative value, and the related synthetic procedures are simple. Sulfonation of bio- and hydro-chars deriving from thermal (e.g., pyrolysis) and hydrothermal processes fully meet these requirements [82]. Their use for some intensified biomass conversion processes has been recently proposed, including the synthesis of 5-HMF [83], furfural [84], levulinic acid [85], alkyl levulinates [86], and other esters [87]. The available data for most of these mild reactions are promising and certainly deserve further research and development.

5. Prefer Efficient Heating Systems

In order to study and optimize these hydrothermal reactions, microwave heating is certainly one of the best choices, being more rapid, energy-saving, and cleaner than the traditional ones, thus suggesting that such new heating systems could lead to more compact factories in the future [21]. In comparison with conventional heating, microwave irradiation has remarkable advantages, such as fast heat transfer and short reaction time, selective and uniform volumetric heating performance, easy operation, high energy efficiency, and reduced formation of by-products, especially in the presence of highly reactive intermediates [88]. In addition to microwaves, also ultrasounds have been proposed for the development of more sustainable and intensified biomass conversion strategies, advantageously exploiting their cavitation effects on many reaction systems. Up to now, this heating technology has been mainly applied for performing mild biomass pretreatments, aimed at the intensified recovery of reducing sugars [89,90], lipids [91], and lignin [92], also in combination with deep eutectic solvents [93], or ionic liquids [94], thus achieving a more effective fractionation of the starting biomass in its components. The efficient utilization of both microwave and ultrasound energy is expected to improve significantly the product yield, efficiency, and environmental friendliness of biomass fractionation processes.

6. Enhance the Concentration of the Target Product

Generally, hydrothermal processes benefit from the dilution of reagents, intermediates, and products, achieving better control of the cross-reactions of the involved species, e.g., a higher selectivity towards the target product. However, a low concentration of the target product would cause too high separation costs in the work-up procedures and, for this reason, it is highly desirable to enhance its final concentration in the reaction mixture, developing the *high-gravity* concept [95]. This goal can be partially achieved by increasing the loading of the starting feedstock, as the solid/liquid ratio, which positively impacts on the environmental performances of the process [31]. However, this choice is not unconditionally beneficial beyond a certain biomass loading, favoring the excessive occurrence of unwanted cross-reactions. For example, in the hydrothermal process for the production of levulinic acid, the biomass loading is limited to 10–20 wt.% [96], whilst a higher one should give practical problems of liquid recovery and promote the excessive formation of undesired solid humins [69]. On the other hand, lower biomass loadings should make the hydrothermal processing economically unviable, due to the high capital investment, power consumption, and heat loss. Therefore, achieving operational status with a high solids loading is still a cumbersome task [97], despite that some practical solutions have been proposed, such as using the pre-hydrolyzed feedstock and starch gels with cement pumps [98].

On this basis, it is crucial to tune properly both the catalytic performances (yield and selectivity) and the final concentration of the target product, preferring the improvement of the latter. A smart improvement of the product concentration could be achieved by performing sequential treatments of recovered reaction mixtures with a new batch of feedstock. In the case of levulinic acid production, this choice should allow achieving higher concentrations of this bio-product (~100 g/L) than those obtained with only one batch experiment (<30 g/L) [96]. Further, in this case, it is necessary to balance the product concentration and yield, the latter worsening excessively at very high concentrations of levulinic acid. As previously stated for the reactive furans, another smart solution provides the use of a [water-organic solvent] biphasic system, which stabilizes the reactive intermediates in the liquid phase, improving the next LA production [99] and, if the organic solvent results as immiscible with water, also allowing LA simultaneous extraction [100]. Alternatively, also the use of an uncommon biphasic system [water-paraffin oil] has been recently proposed, where the latter is used as a non-solvent for the compound of interest. This approach allows an increase in the levulinic acid concentration in the water phase, also in this case with concentrations higher than 100 g/L, leaving enough liquid phase to sustain the processability of a high loading slurry but reducing the water volume to be processed downstream. Therefore, the target product can be easily recovered from the aqueous phase, whilst the organic non-solvent can be advantageously recycled and reused, given its good thermochemical stability [101].

7. Prefer Cascade over Stepwise Reactions

The synthesis of bio-products often involves multistep reactions, which can be carried out (i) in a stepwise manner, e.g., separating and purifying the product from the reaction mixture before performing the subsequent reaction or, more advantageously, (ii) in cascade, thus directly using the intermediates deriving from the previous step to give subsequent reactions, without their further isolation. The stepwise approach is widely used due to incompatible reaction conditions between steps and the poor catalytic specificity and selectivity of the catalysts. Instead, cascade reactions occur directly, avoiding the isolation and purification of synthetic intermediates, and greatly simplifying the operational procedure. The one-pot cascade approach is advantageously related to atom economy, process time, labor and resource management, and waste generation [102], but requires careful tuning of the catalyst properties. Such an approach is particularly attractive when a high selectivity to the target product is desired and, for this purpose, the development of multifunctional catalytic systems, with well-tuned chemical properties, is essential. Moreover, a worsening of the final target product yield may occur, depending on the number and complexity of the involved steps, making this choice extremely attractive, especially when easy conversion steps are required. In this context, a wide and very hot topic is the improvement of mild biomass conversion processes, by developing new catalysts with bifunctional Brønsted/Lewis acid–base properties, which should be tunable in character and strength [103]. Many noteworthy examples are available from the literature, including 5-HMF production through glucose to fructose isomerization, occurring over a solid base/Lewis acid, and the subsequent fructose dehydration over Brønsted acid sites [104]; the following levulinic acid production by harsher Brønsted acid-catalyzed hydrolysis of 5-HMF, but preferably carried out in the cascade approach, starting from carbohydrate precursors [105]; the production of polyols or alkanes via hydrogenation of glucose to sorbitol over metal catalysts and the subsequent hydrogenolysis over metal-acid bifunctional catalysts [106]; the production of isosorbide from C6 carbohydrates via hydrogenation and dehydration, catalyzed by metal-promoted solid acids [107]; the γ-valerolactone production from xylose via Brønsted acid catalysis coupled with Lewis acid- or base-catalyzed Meerwein–Ponndorf–Verley hydrogen transfer, in the presence of isopropanol as the hydrogen donor [108]; the synthesis of 2-methyltetrahydrofuran by bimetallic-catalyzed hydrogenation of γ-valerolactone or, even better, levulinic acid [109]; and the condensation/oligomerization of C5/C6-derived bio-products to longer carbon-chain chemicals and their oxygen removal to give liquid alkanes, by hydrogenation [110]. Considering the multiple processes involved, issues of catalyst stability and economic feasibility remain essential priorities for the next development of the process intensification.

8. Conclusions

Process intensification is a tool that can be very helpful for the development of low-cost processes, with better use of physical spaces and low energy requirements. Chemical, technological, economic, environmental, and regulatory questions have to be considered for improving the solvothermal process intensification. Complete process integration is necessary to get significant cost advantages, preferring the use of cheap and waste biomasses, direct conversion strategies, preferably in cascade, improving the final concentration of the target product, rather than its yield (however maintaining an appreciable yield), in the case of waste starting materials, and exploiting more efficient heating systems, such as microwaves and/or sonication. The development of efficient and cheap heterogeneous catalysts starting from low-cost or negative value precursors, preferably waste produced within the same biomass biorefinery, still requires further research and development, whilst significant progress has been done on the design of multifunctional catalysts for performing cascade reactions. Technological limitations due to the purification of the product/recovery of the acid catalyst have been partially overcome and do not seem to limit the development of the intensification of these processes. On the other hand, environmental problems due to wastewater treatments and legal restrictions of certain types of waste biomasses may slow down their development. Besides, simultaneously with the optimization of the catalysis issue, other strategic business drivers should be taken into account for evaluating the real feasibility of the biomass conversion process, such as biomass transportation cost, agronomic parameters (productivity on a dry basis, input degree), and plant production capacity, thus highlighting that an interdisciplinary life cycle sustainability assessment should be carefully performed at an advanced stage of process development. An accurate evaluation of project feasibility, capital and operating costs, revenues, and profitability measures is imperative, thus helping to bridge the uncertainty associated with a lack of data on investments on larger-scale plants.

Anna Maria Raspolli Galletti
Full Professor
DCCI* - University of Pisa

Claudia Antonetti
Associate Professor
DCCI* - University of Pisa

Domenico Licursi
Post-Doc Fellow
DCCI* - University of Pisa

Author Information (* DCCI: Department of Chemistry and Industrial Chemistry).

Author Contributions: All the authors wrote, revised and supervised the manuscript. All authors have read and agreed to the published version of the manuscript.

Funding: This research received no external funding.

Conflicts of Interest: The authors declare no conflict of interest

References

1. Artz, J.; Palkovits, R. Cellulose-based platform chemical: The path to application. *Curr. Opin. Green Sustain. Chem.* **2018**, *14*, 14–18. [CrossRef]
2. Sadula, S.; Athaley, A.; Zheng, W.; Ierapetritou, M.; Saha, B. Process intensification for cellulosic biorefineries. *ChemSusChem* **2017**, *10*, 2566–2572. [CrossRef] [PubMed]
3. Lachos-Perez, D.; Brown, A.B.; Mudhoo, A.; Martinez, J.; Timko, M.T.; Rostagno, M.A.; Forster-Carneiro, T. Applications of subcritical and supercritical water conditions for extraction, hydrolysis, gasification, and carbonization of biomass: A critical review. *Biofuel Res. J.* **2017**, *14*, 611–626. [CrossRef]
4. Kumar, M.; Oyedun, A.O.; Kumar, A. A review on the current status of various hydrothermal technologies on biomass feedstock. *Renew. Sustain. Energy Rev.* **2018**, *81*, 1742–1770. [CrossRef]
5. Xu, C.; Paone, E.; Rodríguez-Padrón, D.; Luque, R.; Mauriello, F. Recent catalytic routes for the preparation and the upgrading of biomass derived furfural and 5-hydroxymethylfurfural. *Chem. Soc. Rev.* **2020**, *49*, 4273–4306. [CrossRef]
6. Kang, S.; Fu, J.; Zhang, G. From lignocellulosic biomass to levulinic acid: A review on acid-catalyzed hydrolysis. *Renew. Sustain. Energy Rev.* **2018**, *94*, 340–362. [CrossRef]
7. Rachamontree, P.; Douzou, T.; Cheenkachorn, K.; Sriariyanun, M.; Rattanaporn, K. Furfural: A sustainable platform chemical and fuel. *Appl. Sci. Eng. Prog.* **2020**, *13*, 3–10. [CrossRef]
8. Metkar, P.S.; Till, E.J.; Corbin, D.R.; Pereira, C.J.; Hutchenson, K.W.; Sengupta, S.K. Reactive distillation process for the production of furfural using solid acid catalysts. *Green Chem.* **2015**, *17*, 1453–1466. [CrossRef]
9. Thoma, C.; Konnerth, J.; Sailer-Kronlachner, W.; Solt, P.; Rosenau, T.; van Herwijnen, H.W.G. Current situation of the challenging scale-up development of hydroxymethylfurfural production. *ChemSusChem* **2020**, *13*, 3544–3564. [CrossRef]
10. Yan, P.; Xia, M.; Chen, S.; Han, W.; Wang, H.; Zhu, W. Unlocking biomass energy: Continuous high-yield production of 5-hydroxymethylfurfural in water. *Green Chem.* **2020**. [CrossRef]
11. AVA Biochem: Bio-Based Chemistry. Available online: http://www.ava-biochem.com (accessed on 1 August 2020).
12. Bello, S.; Salim, I.; Méndez-Trelles, P.; Rodil, E.; Feijoo, G.; Moreira, M.T. Environmental sustainability assessment of HMF and FDCA production from lignocellulosic biomass through life cycle assessment (LCA). *Holzforschung* **2018**, *73*, 105–115. [CrossRef]
13. Bello, S.; Méndez-Trelles, P.; Rodil, E.; Feijoo, G.; Moreira, M.T. Towards improving the sustainability of bioplastics: Process modelling and life cycle assessment of two separation routes for 2,5-furandicarboxylic acid. *Sep. Purif. Technol.* **2020**, *233*, 116056. [CrossRef]
14. Sousa, A.F.; Vilela, C.; Fonseca, A.C.; Matos, M.; Freire, C.S.R.; Gruter, G.J.M.; Coelho, J.F.J.; Silvestre, A.J.D. Biobased polyesters and other polymers from 2,5-furandicarboxylic acid: A tribute to furan excellency. *Polym. Chem.* **2015**, *6*, 5961–5983. [CrossRef]

15. Meramo-Hurtado, S.I.; Ojeda, K.A.; Sanchez-Tuiran, E. Environmental and safety assessments of industrial production of levulinic acid via acid-catalyzed dehydration. *ACS Omega* **2019**, *4*, 22302–22312. [CrossRef] [PubMed]
16. Mika, L.T.; Cséfalvay, E.; Németh, Á. Catalytic conversion of carbohydrates to initial platform chemicals: Chemistry and sustainability. *Chem. Rev.* **2018**, *118*, 505–613. [CrossRef]
17. Pileidis, F.D.; Titirici, M.-M. Levulinic acid biorefineries: New challenges for efficient utilization of biomass. *ChemSusChem* **2016**, *9*, 562–582. [CrossRef]
18. Li, C.; Duan, C.; Fang, J.; Li, H. Process intensification and energy saving of reactive distillation for production of ester compounds. *Chin. J. Chem. Eng.* **2019**, *27*, 1307–1323. [CrossRef]
19. Silva, J.F.L.; Grekin, R.; Mariano, A.P.; Filho, R.M. Making levulinic acid and ethyl levulinate economically viable: A worldwide techno-economic and environmental assessment of possible routes. *Energy Technol.* **2018**, *6*, 613–639. [CrossRef]
20. Morone, A.; Apte, M.; Pandey, R.A. Levulinic acid production from renewable waste resources: Bottlenecks, potential remedies, advancements and applications. *Renew. Sustain. Energy Rev.* **2015**, *51*, 548–565. [CrossRef]
21. Antonetti, C.; Bonari, E.; Licursi, D.; Nassi, N.; Raspolli Galletti, A.M. Hydrothermal conversion of giant reed to furfural and levulinic acid: Optimization of the process under microwave irradiation and investigation of distinctive agronomic parameters. *Molecules* **2015**, *20*, 21232–21253. [CrossRef]
22. Rivas, S.; Raspolli Galletti, A.M.; Antonetti, C.; Santos, V.; Parajó, J.C. Sustainable conversion of *Pinus pinaster* wood into biofuel precursors: A biorefinery approach. *Fuel* **2016**, *164*, 51–58. [CrossRef]
23. Di Fidio, N.; Raspolli Galletti, A.M.; Fulignati, S.; Licursi, D.; Liuzzi, F.; De Bari, I.; Antonetti, C. Multi-step exploitation of raw arundo donax L. for the selective synthesis of second-generation sugars by chemical and biological route. *Catalysts* **2020**, *10*, 79. [CrossRef]
24. Signoretto, M.; Taghavi, S.; Ghedini, E.; Menegazzo, F. Catalytic production of levulinic Acid (LA) from actual biomass. *Molecules* **2019**, *24*, 2760. [CrossRef] [PubMed]
25. Woo, H.; Acuna, M.; Moroni, M.; Taskhiri, M.S.; Turner, P. Optimizing the location of biomass energy facilities by integrating multi-criteria analysis (MCA) and geographical information systems (GIS). *Forests* **2018**, *9*, 585. [CrossRef]
26. Khoo, H.H.; Eufrasio-Espinosa, R.M.; Koh, L.S.C.; Sharratt, P.N.; Isoni, V. Sustainability assessment of biorefinery production chains: A combined LCA-supply chain approach. *J. Clean. Prod.* **2019**, *235*, 1116–1137. [CrossRef]
27. Lui, M.Y.; Wong, C.Y.Y.; Choi, A.W.-T.; Mui, Y.F.; Qi, L.; Horváth, I.T. Valorization of carbohydrates of agricultural residues and food wastes: A key strategy for carbon conservation. *ACS Sustain. Chem. Eng.* **2019**, *7*, 17799–17807. [CrossRef]
28. Licursi, D.; Antonetti, C.; Martinelli, M.; Ribechini, E.; Zanaboni, M.; Raspolli Galletti, A.M. Monitoring/characterization of stickies contaminants coming from a papermaking plant—Toward an innovative exploitation of the screen rejects to levulinic acid. *Waste Manag.* **2016**, *49*, 469–482. [CrossRef]
29. Chen, W.-T.; Haque, A.; Lu, T.; Aierzhati, A.; Reimonn, G. A perspective on hydrothermal processing of sewage sludge. *Curr. Opin. Environ. Sci. Health* **2020**, *14*, 63–73. [CrossRef]
30. Reißmann, D.; Thrän, D.; Bezama, A. Hydrothermal processes as treatment paths for biogenic residues in Germany: A review of the technology, sustainability and legal aspects. *J. Clean. Prod.* **2018**, *172*, 239–252. [CrossRef]
31. Reißmann, D.; Thrän, D.; Bezama, A. Key development factors of hydrothermal processes in Germany by 2030: A fuzzy logic analysis. *Energies* **2018**, *11*, 3532. [CrossRef]
32. Dulie, N.W.; Woldeyes, B.; Demsash, H.D.; Jabasingh, A.S. An insight into the valorization of hemicellulose fraction of biomass into furfural: Catalytic conversion and product separation. *Waste Biomass Valor.* **2020**. [CrossRef]
33. Inyang, V.M.; Lokhat, D. Separation of carboxylic acids: Conventional and intensified processes and effects of process engineering parameters. In *Valorization of Biomass to Value-Added Commodities*; Daramola, M.O., Ayeni, A.O., Eds.; Springer Nature: Cham, Switzerland, 2020; pp. 469–506.
34. Alcocer-García, H.; Segovia-Hernández, J.G.; Prado-Rubio, O.A.; Sánchez-Ramírez, E.; Quiroz-Ramírez, J.J. Multi-objective optimization of intensified processes for the purification of levulinic acid involving economic and environmental objectives. *Chem. Eng. Process.* **2019**, *136*, 123–137. [CrossRef]

35. Alcocer-García, H.; Segovia-Hernández, J.G.; Prado-Rubio, O.A.; Sánchez-Ramírez, E.; Quiroz-Ramírez, J.J. Multi-objective optimization of intensified processes for the purification of levulinic acid involving economic and environmental objectives. Part II: A comparative study of dynamic properties. *Chem. Eng. Process.* **2020**, *147*, 107745. [CrossRef]
36. Möller, M.; Nilges, P.; Harnisch, F.; Schröder, U. Subcritical water as reaction environment: Fundamentals of hydrothermal biomass transformation. *ChemSusChem* **2011**, *4*, 566–579. [CrossRef]
37. Bouxin, F.; Marinkovic, S.; Le Bras, J.; Estrine, B. Direct conversion of xylan into alkyl pentosides. *Carbohydr. Res.* **2010**, *345*, 2469–2473. [CrossRef]
38. Puga, A.V.; Corma, A. Direct conversion of cellulose into alkyl glycoside surfactants. *ChemistrySelect* **2017**, *2*, 2495–2498. [CrossRef]
39. Alipour, S.; Omidvarborna, H.; Kim, D.-S. A review on synthesis of alkoxymethyl furfural, a biofuel candidate. *Renew. Sustain. Energy Rev.* **2017**, *71*, 908–926. [CrossRef]
40. Démolis, A.; Essayem, N.; Rataboul, F. Synthesis and applications of alkyl levulinates. *ACS Sustain. Chem. Eng.* **2014**, *2*, 1338–1352. [CrossRef]
41. Zhang, Q.; Hu, J.; Lee, D.-J. Pretreatment of biomass using ionic liquids: Research updates. *Renew. Energy* **2017**, *111*, 77–84. [CrossRef]
42. Tan, Y.T.; Chua, A.S.M.; Ngoh, G.C. Deep eutectic solvent for lignocellulosic biomass fractionation and the subsequent conversion to bio-based products—A review. *Bioresour. Technol.* **2020**, *297*, 122522. [CrossRef]
43. Chen, Y.; Mu, T. Application of deep eutectic solvents in biomass pretreatment and conversion. *Green Energy Environ.* **2019**, *4*, 95–115. [CrossRef]
44. Abbott, A.P.; Bell, T.J.; Handa, S.; Stoddart, B. O-Acetylation of cellulose and monosaccharides using a zinc based ionic liquid. *Green Chem.* **2005**, *7*, 705–707. [CrossRef]
45. Abbott, A.P.; Bell, T.J.; Handa, S.; Stoddart, B. Cationic functionalisation of cellulose using a choline based ionic liquid analogue. *Green Chem.* **2006**, *8*, 784–786. [CrossRef]
46. Selkälä, T.; Sirviö, J.A.; Lorite, G.S.; Liimatainen, H. Anionically stabilized cellulose nanofibrils through succinylation pretreatment in urea-lithium chloride deep eutectic solvent. *ChemSusChem* **2016**, *9*, 3074–3083. [CrossRef] [PubMed]
47. Esteban, J.; Vorholt, A.J.; Leitner, W. An overview of the biphasic dehydration of sugars to 5-hydroxymethylfurfural and furfural: A rational selection of solvents using COSMO-RS and selection guides. *Green Chem.* **2020**, *22*, 2097–2128. [CrossRef]
48. Liu, F.; Xue, Z.; Zhao, X.; Mou, H.; He, J.; Mu, T. Catalytic deep eutectic solvents for highly efficient conversion of cellulose to gluconic acid with gluconic acid self-precipitation separation. *Chem. Commun.* **2018**, *54*, 6140–6143. [CrossRef]
49. Park, C.; Lee, J. Recent achievements in CO_2-assisted and CO_2-catalyzed biomass conversion reactions. *Green Chem.* **2020**, *22*, 2628–2642. [CrossRef]
50. Wu, K.; Wu, Y.; Chen, Y.; Chen, H.; Wang, J.; Yang, M. Heterogeneous catalytic conversion of biobased chemicals into liquid fuels in the aqueous phase. *ChemSusChem* **2016**, *9*, 1355–1385. [CrossRef]
51. De, S.; Dutta, S.; Saha, B. Critical design of heterogeneous catalysts for biomass valorization: Current thrust and emerging prospects. *Catal. Sci. Technol.* **2016**, *6*, 7364–7385. [CrossRef]
52. Xia, Q.; Wang, Y. Niobium-based catalysts for biomass conversion. In *Nanoporous Catalysts for Biomass Conversion*; Xiao, F.-S., Wang, L., Stevens, C.V., Eds.; John Wiley & Sons Ltd.: Oxford, UK, 2018; pp. 253–282.
53. Rackemann, D.W.; Doherty, W.O.S. The conversion of lignocellulosics to levulinic acid. *Biofuels Bioprod. Biorefin.* **2011**, *5*, 198–214. [CrossRef]
54. Mukherjee, A.; Dumont, M.-J.; Raghavan, V. Review: Sustainable production of hydroxymethylfurfural and levulinic acid: Challenges and opportunities. *Biomass Bioenergy* **2015**, *72*, 143–183. [CrossRef]
55. Thoresen, P.P.; Matsakas, L.; Rova, U.; Christakopoulos, P. Recent advances in organosolv fractionation: Towards biomass fractionation technology of the future. *Bioresour. Technol.* **2020**, *306*, 123189. [CrossRef] [PubMed]
56. Zang, G.; Shah, A.; Wan, C. Techno-economic analysis of an integrated biorefinery strategy based on one-pot biomass fractionation and furfural production. *J. Clean. Prod.* **2020**, *260*, 120837. [CrossRef]
57. Rivas, S.; Raspolli Galletti, A.M.; Antonetti, C.; Licursi, D.; Santos, V.; Parajó, J.C. A biorefinery cascade conversion of hemicellulose-free *Eucalyptus globulus* wood: Production of concentrated levulinic acid solutions for γ-valerolactone sustainable preparation. *Catalysts* **2018**, *8*, 169. [CrossRef]

58. Licursi, D.; Antonetti, C.; Bernardini, J.; Cinelli, P.; Coltelli, M.B.; Lazzeri, A.; Martinelli, M.; Raspolli Galletti, A.M. Characterization of the *Arundo Donax* L. solid residue from hydrothermal conversion: Comparison with technical lignins and application perspectives. *Ind. Crop. Prod.* **2015**, *76*, 1008–1024. [CrossRef]
59. Licursi, D.; Antonetti, C.; Fulignati, S.; Vitolo, S.; Puccini, M.; Ribechini, E.; Bernazzani, L.; Raspolli Galletti, A.M. In-depth characterization of valuable char obtained from hydrothermal conversion of hazelnut shells to levulinic acid. *Bioresour. Technol.* **2017**, *244*, 880–888. [CrossRef]
60. Bernardini, J.; Licursi, D.; Anguillesi, I.; Cinelli, P.; Coltelli, M.-B.; Antonetti, C.; Raspolli Galletti, A.M.; Lazzeri, A. Exploitation of *Arundo donax* L. hydrolysis residue for the green synthesis of flexible polyurethane foams. *Biol. Res.* **2017**, *12*, 3630–3655. [CrossRef]
61. Zhang, Z.; Zhu, Z.; Shen, B.; Liu, L. Insights into biochar and hydrochar production and applications: A review. *Energy* **2019**, *171*, 581–598. [CrossRef]
62. Licursi, D.; Antonetti, C.; Mattonai, M.; Pérez-Armada, L.; Rivas, S.; Ribechini, E.; Raspolli Galletti, A.M. Multi-valorisation of giant reed (*Arundo Donax* L.) to give levulinic acid and valuable phenolic antioxidants. *Ind. Crop. Prod.* **2018**, *112*, 6–17. [CrossRef]
63. Ferreira, J.A.; Taherzadeh, M.J. Improving the economy of lignocellulose-based biorefineries with organosolv pretreatment. *Bioresour. Technol.* **2020**, *299*, 122695. [CrossRef]
64. Bajwa, D.S.; Pourhashem, G.; Ullah, A.H.; Bajwa, S.G. A concise review of current lignin production, applications, products and their environmental impact. *Ind. Crop. Prod.* **2019**, *139*, 111526. [CrossRef]
65. Shi, N.; Liu, Q.; Ju, R.; He, X.; Zhang, Y.; Tang, S.; Ma, L. Condensation of α-carbonyl aldehydes leads to the formation of solid humins during the hydrothermal degradation of carbohydrates. *ACS Omega* **2019**, *4*, 7330–7343. [CrossRef] [PubMed]
66. Shen, H.; Shan, H.; Liu, L. Evolution process and controlled synthesis of humins with 5-hydroxymethylfurfural (HMF) as model molecule. *ChemSusChem* **2020**, *13*, 513–519. [CrossRef] [PubMed]
67. Shi, N.; Liu, Q.; Cen, H.; Ju, R.; He, X.; Ma, L. Formation of humins during degradation of carbohydrates and furfural derivatives in various solvents. *Biomass Conv. Biorefin.* **2020**, *10*, 277–287. [CrossRef]
68. Cheng, B.; Wang, X.; Lin, Q.; Zhang, X.; Meng, L.; Sun, R.-C.; Xin, F.; Ren, J. New understandings of the relationship and initial formation mechanism for pseudo-lignin, humins, and acid-induced hydrothermal carbon. *J. Agric. Food Chem.* **2018**, *66*, 11981–11989. [CrossRef]
69. Lopes, E.S.; Silva, J.F.L.; Rivera, E.C.; Gomes, A.P.; Lopes, M.S.; Filho, R.M.; Tovar, L.P. Challenges to levulinic acid and humins valuation in the sugarcane bagasse biorefinery concept. *Biol. Energy Res.* **2020**. [CrossRef]
70. Shen, Y. A review on hydrothermal carbonization of biomass and plastic wastes to energy products. *Biomass Bioenergy* **2020**, *134*, 105479. [CrossRef]
71. Zhang, X.; Wang, Y.; Cai, J.; Wilson, K.; Lee, A.F. Bio/hydrochar sorbents for environmental remediation. *Energy Environ. Mater.* **2020**. [CrossRef]
72. Pin, J.-M.; Guigo, N.; Mija, A.; Vincent, L.; Sbirrazzuoli, N.; van der Waal, J.C.; de Jong, E. Valorization of biorefinery side-stream products: Combination of humins with polyfurfuryl alcohol for composite elaboration. *ACS Sustain. Chem. Eng.* **2014**, *2*, 2182–2190. [CrossRef]
73. Kang, S.; Fu, J.; Zhang, G.; Zhang, W.; Yin, H.; Xu, Y. Synthesis of humin-phenol-formaldehyde adhesive. *Polymers* **2017**, *9*, 373. [CrossRef]
74. Sangregorio, A.; Muralidhara, A.; Guigo, N.; Thygesen, L.G.; Marlair, G.; Angelici, C.; de Jong, E.; Sbirrazzuoli, N. Humin based resin for wood modification and property improvement. *Green Chem.* **2020**, *22*, 2786–2798. [CrossRef]
75. Hoang, T.M.C.; van Eck, E.R.H.; Bula, W.P.; Gardeniers, J.G.E.; Lefferts, L.; Seshan, K. Humin based by-products from biomass processing as a potential carbonaceous source for synthesis gas production. *Green Chem.* **2015**, *17*, 959–972. [CrossRef]
76. Kang, S.; Zhang, G.; Yang, Q.; Tu, J.; Guo, X.; Qin, F.G.F.; Xu, Y. A new technology for utilization of biomass hydrolysis residual humins for acetic acid production. *Biol. Res.* **2016**, *11*, 9496–9505. [CrossRef]
77. Maerten, S.G.; Voß, D.; Liauw, M.A.; Albert, J. Selective catalytic oxidation of humins to low-chain carboxylic acids with tailor-made polyoxometalate catalysts. *ChemistrySelect* **2017**, *2*, 7296–7302. [CrossRef]
78. YXY - Avantium. Available online: https://www.avantium.com/technologies/yxy/ (accessed on 1 August 2020).

79. Eerhart, A.J.J.E.; Huijgen, W.J.J.; Grisel, R.J.H.; van der Waal, J.C.; de Jong, E.; de Sousa Dias, A.; Faaij, A.P.C.; Patel, M.K. Fuels and plastics from lignocellulosic biomass via the furan pathway; a technical analysis. *RSC Adv.* **2014**, *4*, 3536–3549. [CrossRef]
80. Wang, Y.; Agarwal, S.; Kloekhorst, A.; Heeres, H.J. Catalytic hydrotreatment of humins in mixtures of formic acid/2-propanol with supported ruthenium catalysts. *ChemSusChem* **2016**, *9*, 951–961. [CrossRef]
81. Sun, J.; Cheng, H.; Zhang, Y.; Zhang, Y.; Lan, X.; Zhang, Y.; Xia, Q.; Ding, D. Catalytic hydrotreatment of humins into cyclic hydrocarbons over solid acid supported metal catalysts in cyclohexane. *J. Energy Chem.* **2020**, *53*, 329–339. [CrossRef]
82. Zhong, R.; Sels, B.F. Sulfonated mesoporous carbon and silica-carbon nanocomposites for biomass conversion. *Appl. Catal. B Environ.* **2018**, *236*, 518–545. [CrossRef]
83. Perez, G.P.; Dumont, M.-J. Production of HMF in high yield using a low cost and recyclable carbonaceous catalyst. *Chem. Eng. J.* **2020**, *382*, 122766. [CrossRef]
84. Wang, Y.; Delbecq, F.; Kwapinski, W.; Len, C. Application of sulfonated carbon-based catalyst for the furfural production from d-xylose and xylan in a microwave-assisted biphasic reaction. *Mol. Catal.* **2017**, *438*, 167–172. [CrossRef]
85. Wang, K.; Jiang, J.; Liang, X.; Wu, H.; Xu, J. Direct conversion of cellulose to levulinic acid over multifunctional sulfonated humins in sulfolane–water solution. *ACS Sustain. Chem. Eng.* **2018**, *6*, 15092–15099. [CrossRef]
86. Pileidis, F.D.; Tabassum, M.; Coutts, S.; Titirici, M.-M. Esterification of levulinic acid into ethyl levulinate catalysed by sulfonated hydrothermal carbons. *Chin. J. Catal.* **2014**, *35*, 929–936. [CrossRef]
87. Yang, Y.-T.; Yang, X.-X.; Wang, Y.-T.; Luo, J.; Zhang, F.; Yang, W.-J.; Chen, J.-H. Alcohothermal carbonization of biomass to prepare novel solid catalysts for oleic acid esterification. *Fuel* **2018**, *219*, 166–175. [CrossRef]
88. Martín, A.; Navarrete, A. Microwave-assisted process intensification techniques. *Curr. Opin. Green Sustain. Chem.* **2018**, *11*, 70–75. [CrossRef]
89. Joshi, S.M.; Gogate, P.R. Intensification of dilute acid hydrolysis of spent tea powder using ultrasound for enhanced production of reducing sugars. *Ultrason. Sonochem.* **2020**, 104843. [CrossRef]
90. Zhang, Y.; Li, T.; Shen, Y.; Wang, L.; Zhang, H.; Qi, H.; Qi, X. Extrusion followed by ultrasound as a chemical-free pretreatment method to enhance enzymatic hydrolysis of rice hull for fermentable sugars production. *Ind. Crop. Prod.* **2020**, *149*, 112356. [CrossRef]
91. Ma, G.; Mu, R.; Capareda, S.C.; Qi, F. Use of ultrasound for aiding lipid extraction and biodiesel production of microalgae harvested by chitosan. *Environ. Technol.* **2020**. [CrossRef]
92. Subhedar, P.B.; Ray, P.; Gogate, P.R. Intensification of delignification and subsequent hydrolysis for the fermentable sugar production from lignocellulosic biomass using ultrasonic irradiation. *Ultrason. Sonochem.* **2018**, *40*, 140–150. [CrossRef]
93. Cherif, M.M.; Grigorakis, S.; Halahlah, A.; Loupassaki, S.; Makris, D.P. High-efficiency extraction of phenolics from wheat waste biomass (bran) by combining deep eutectic solvent, ultrasound-assisted pretreatment and thermal treatment. *Environ. Process.* **2020**, *7*, 845–859. [CrossRef]
94. Miranda, R.C.M.; Neta, J.V.; Ferreira, L.F.R.; Júnior, W.A.G.; do Nascimento, C.S.; Gomes, E.B.; Mattedi, S.; Soares, C.M.F.; Lima, A.S. Pineapple crown delignification using low-cost ionic liquid based on ethanolamine and organic acids. *Carbohydr. Polym.* **2019**, *206*, 302–308. [CrossRef]
95. Xiros, C.; Janssen, M.; Byström, R.; Børresen, B.T.; Cannella, D.; Jørgensen, H.; Koppram, R.; Larsson, C.; Olsson, L.; Tillman, A.M.; et al. Toward a sustainable biorefinery using high-gravity technology. *Biofuels Bioprod. Biorefin.* **2017**, *11*, 15–27. [CrossRef]
96. Kang, S.; Yu, J. An intensified reaction technology for high levulinic acid concentration from lignocellulosic biomass. *Biomass Bioenergy* **2016**, *95*, 214–220. [CrossRef]
97. Valdivia, M.; Galan, J.L.; Laffarga, J.; Ramos, J.-L. Biofuels 2020: Biorefineries based on lignocellulosic materials. *Microb. Biotechnol.* **2016**, *9*, 585–594. [CrossRef] [PubMed]
98. Peterson, A.A.; Vogel, F.; Lachance, R.P.; Fröling, M.; Antal, M.J.; Tester, J.W. Thermochemical biofuel production in hydrothermal media: A review of sub- and supercritical water technologies. *Energy Environ. Sci.* **2008**, *1*, 32–65. [CrossRef]
99. Dutta, S.; Yu, I.K.M.; Tsang, D.C.W.; Su, Z.; Hu, C.; Wu, K.C.W.; Yip, A.C.K.; Ok, Y.S.; Poon, C.S. Influence of green solvent on levulinic acid production from lignocellulosic paper waste. *Bioresour. Technol.* **2020**, *298*, 122544. [CrossRef]

100. Bokade, V.; Moondra, H.; Niphadkar, P. Highly active Brønsted acidic silicon phosphate catalyst for direct conversion of glucose to levulinic acid in MIBK–water biphasic system. *SN Appl. Sci.* **2020**, *2*. [CrossRef]
101. Licursi, D.; Antonetti, C.; Parton, R.; Raspolli Galletti, A.M. A novel approach to biphasic strategy for intensification of the hydrothermal process to give levulinic acid: Use of an organic non-solvent. *Bioresour. Technol.* **2018**, *264*, 180–189. [CrossRef]
102. Liu, Y.; Nie, Y.; Lu, X.; Zhang, X.; He, H.; Pan, F.; Zhou, L.; Liu, X.; Ji, X.; Zhang, S. Cascade utilization of lignocellulosic biomass to high-value products. *Green Chem.* **2019**, *21*, 3499–3535. [CrossRef]
103. Wang, Z.; Huang, J. Brønsted-Lewis acids for efficient conversion of renewables. In *Production of Biofuels and Chemicals with Bifunctional Catalysts*; Fang, Z., Smith, R.L., Li, H., Eds.; Springer Nature: Singapore, 2017; pp. 99–136. [CrossRef]
104. Marianou, A.A.; Michailof, C.M.; Pineda, A.; Iliopoulou, E.F.; Triantafyllidis, K.S.; Lappas, A.A. Effect of Lewis and Brønsted acidity on glucose conversion to 5-HMF and lactic acid in aqueous and organic media. *Appl. Catal. A Gen.* **2018**, *555*, 75–87. [CrossRef]
105. Boonyakarn, T.; Wataniyakul, P.; Boonnoun, P.; Quitain, A.T.; Kida, T.; Sasaki, M.; Laosiripojana, N.; Jongsomjit, B.; Shotipruk, A. Enhanced levulinic acid production from cellulose by combined Brønsted hydrothermal carbon and Lewis acid catalysts. *Ind. Eng. Chem. Res.* **2019**, *58*, 2697–2703. [CrossRef]
106. Vilcocq, L.; Cabiac, A.; Especel, C.; Lacombe, S.; Duprez, D. New insights into the mechanism of sorbitol transformation over an original bifunctional catalytic system. *J. Catal.* **2014**, *320*, 16–25. [CrossRef]
107. Dussenne, C.; Delaunay, T.; Wiatz, V.; Wyart, H.; Suisse, I.; Sauthier, M. Synthesis of isosorbide: An overview of challenging reactions. *Green Chem.* **2017**, *19*, 5332–5344. [CrossRef]
108. Yamaguchi, A.; Sato, O.; Mimura, N.; Shirai, M. One-pot conversion of cellulose to isosorbide using supported metal catalysts and ion-exchange resin. *Catal. Commun.* **2015**, *67*, 59–63. [CrossRef]
109. Licursi, D.; Antonetti, C.; Fulignati, S.; Giannoni, M.; Raspolli Galletti, A.M. Cascade strategy for the tunable catalytic valorization of levulinic acid and γ-valerolactone to 2-methyltetrahydrofuran and alcohols. *Catalysts* **2018**, *8*, 277. [CrossRef]
110. Kaldstro, M.; Lindblad, M.; Lamminpää, K.; Wallenius, S.; Toppinen, S. Carbon chain length increase reactions of platform molecules derived from C5 and C6 sugars. *Ind. Eng. Chem. Res.* **2017**, *56*, 13356–13366. [CrossRef]

© 2020 by the authors. Licensee MDPI, Basel, Switzerland. This article is an open access article distributed under the terms and conditions of the Creative Commons Attribution (CC BY) license (http://creativecommons.org/licenses/by/4.0/).

Article

Acetylation of Eugenol on Functionalized Mesoporous Aluminosilicates Synthesized from Amazonian *Flint* Kaolin

Alex de Nazaré de Oliveira [1,2,3], Erika Tallyta Leite Lima [1,2], Eloisa Helena de Aguiar Andrade [1,4], José Roberto Zamian [1,2], Geraldo Narciso da Rocha Filho [1,2], Carlos Emmerson Ferreira da Costa [1,2], Luíza Helena de Oliveira Pires [5], Rafael Luque [6,*] and Luís Adriano Santos do Nascimento [1,2,7,*]

1. Graduation Program in Chemistry, Federal University of Pará, Augusto Corrêa Street, Guamá, Belém, PA 66075-110, Brazil; alexoliveiraquimica@hotmail.com (A.d.N.d.O.); erikatallyta@hotmail.com (E.T.L.L.); elena@ufpa.br (E.H.d.A.A.); zamian@ufpa.br (J.R.Z.); geraldonrf@gmail.com (G.N.d.R.F.); emmerson@ufpa.br (C.E.F.d.C.)
2. Laboratory of Oils of the Amazon, Federal University of Pará, Perimetral Avenue, Guamá, Belém, PA 66075-750, Brazil
3. Department of Exact and Technologic Sciences, Federal University of Amapá, Rod. Juscelino Kubitschek, km 02-Jardim Marco Zero, Macapá, AP 68903-419, Brazil
4. Adolpho Ducke Laboratory, Botany Coordinating, Museu Paraense Emílio Goeldi, Perimetral Avenue, Terra Firme, Belém, PA 66077-830, Brazil
5. School of Application, Federal University of Pará, Pará 66077-585, Brazil; lulenapires@hotmail.com
6. Department of Organic Chemistry, Universidad de Córdoba, Ctra Nnal IV-A, Km 396, E14014 Cordoba, Spain
7. Graduation Program in Biotechnology, Federal University of Pará, Augusto Corrêa Street, Guamá, Belém, PA 66075-110, Brazil
* Correspondence: q62alsor@uco.es (R.L.); adrlui1@yahoo.com.br (L.A.S.d.N.); Tel.: +55-91-98171-4947 (R.L.)

Received: 16 March 2020; Accepted: 23 April 2020; Published: 27 April 2020

Abstract: The present work was aimed to investigate the catalytic activity of a mesoporous catalyst synthesized from 3-mercaptopropyltrimethoxysilane (MPTS) functionalized Amazonian *flint* kaolin in the acetylation of eugenol with acetic anhydride. Materials were characterized by thermogravimetry (TGA), N_2 adsorption (BET), X-ray dispersive energy spectroscopy (EDX), X-ray diffraction (XRD), Fourier transform infrared spectroscopy (FTIR) and acid-base titration. The results presented proved the efficiency of *flint* kaolin as an alternative source in the preparation of mesoporous materials, since the material exhibited textural properties (specific surface area of 1071 $m^2\ g^{-1}$, pore volume of 1.05 $cm^3\ g^{-1}$ and pore diameter of 3.85 nm) and structural properties (d_{100} = 4.35 nm, a_0 = 5.06 nm and W_t = 1.21 nm) within the required and characteristic material standards. The catalyst with the total amount of acidic sites of 4.89 mmol $H^+\ g^{-1}$ was efficient in converting 99.9% of eugenol (eugenol to acetic anhydride molar ratio of 1:5, 2% catalyst, temperature and reaction time 80 °C and 40 min reaction). In addition, the reused catalyst could be successfully recycled with 92% conversion activity under identical reaction conditions.

Keywords: eugenol; acetylation; *flint* kaolin; mesoporous aluminosilicate; functionalization; heterogeneous catalysis

1. Introduction

Aedes aegypti is an urban mosquito, which proliferates in areas of greater population density. Because it is typical of tropical and subtropical regions, it is considered a potential transmitter of the chikungunya, Zika virus (diseases that can generate other diseases, such as microcephaly and

Guillain–Barré), in addition to dengue and urban yellow fever. [1–4]. One of the main measures to prevent and combat *Aedes Aegypti* is with the use of insecticides and synthetic larvicides such as organochlorines, organophosphates, and others, to prevent the infestation of the adult and larval mosquitoes [1,3]. However, studies have shown that controlling the larval stage mosquito population is a more effective alternative than insecticide use in adult individuals [2,4–6]. For this reason, several researches are trying to develop new effective natural larvicides and insecticides, without toxic effect on man and the environment [2]. In this context, excellent results against *A. aegypti* larvae using eugenol and eugenyl acetate essential oils have been previously reported in the literature [2,4–6].

Eugenol is a natural product, belonging to the class of phenylpropanoids, found in greater quantity in the essential oil of cloves (*Eugenia Caryophyllata*) and exhibits a wide variety of biological, pharmacological and other applications, being a product of great commercial interest [4,5,7]. The interest of the scientific community for eugenol is justified by the fact that it is a reactant for synthesizing various natural and bioactive products through reactions such as acetylation reaction [4,8,9] or benzoylation [10–12], within the scope of renewable resources. Eugenyl acetate obtained from eugenol, even in small amounts, has been reported as effective against *A. aegypti* mosquito larvae at development stage [2,4,6], being considered an efficient low cost natural larvicide [2–4,6,13].

The synthesis of eugenol esters often proceeds through homogeneous catalysis most often they are toxic or dangerous, such as mineral acids, chlorides, pyridine and their derivatives and more [2,14–16]. In addition to safety considerations, these technologies are also inefficient due to additional waste separation and treatment steps to isolate the product and thus are economically disadvantageous [10–12]. Although biocatalytic processes are alternative and environmentally friendly, to obtain the ester, in many cases it presents low capacity of reuse, making the process costly [5,17,18].

The development of solid catalysts is no longer seen simply in terms of cost and energy efficiency optimization, but as a clean technology [19]. The use of heterogeneous acid catalysts offers advantages over homogeneous ones such as solids recovery and recycling, as well as reducing environmental impacts [20–30]. Given the growing environmental problems, it is highly desirable to use heterogeneous catalysts in place of traditional homogeneous ones (corrosive, toxic and expensive) [10,17,24,31,32].

MCM-41 (Mobil Composition of Matter No. 41) mesoporous molecular sieves have been the most studied member of the M41S (Mobil 41 Synthesis) family. Their physical properties such as high pore volume and thermal stability, uniform pore diameters (from 1.5 to 10 nm) and large surface area [22,26,33–35] allow them to be applied as, heterogeneous catalytic supports and catalysts for several areas [24,28,33–35]. Generally, MCM-41 is synthesized using different sources of silica such as sodium silicate, tetraethylorthosilicate (TEOS), tetramethylammonium silicate (TMA-silicate) and the structure driver (cetyltrimethylammonium bromide, CTABr). However, these silica precursors have some disadvantages such as toxicity and the high cost of producing the material [36,37]. For economic and environmental reasons, the search for a synthesis procedure using alternative sources of silica, such as high purity kaolin, has been intensified aiming to obtain mesoporous material with a well-defined hexagonal arrangement [38–42]. Additionally, the waste from the kaolin processing [28] has proven to be an excellent source of silicon and aluminum for synthesis of highly ordered Al–MCM-41.

The process of transforming raw kaolin to a commercial product generates large amounts of waste that can have a great environmental impact. In this processing, a kaolinitic residue, kaolin *flint*, is generated with a high iron content and disposed in the mine shortly after exploration [25,43,44]. In order to minimize the environmental impacts caused by the residues of the kaolin industrial, our research group has been working on the use of this waste as raw material in the synthesis of catalytic materials and it has been shown to be quite viable because this material is composed essentially of kaolinite, the most common clay used as a source of Si and Al in the synthesis of materials. The use of kaolinitic tailings as raw material for catalyst synthesis [20,21,23,25], catalytic supports [22,27,28] and for mesoporousaluminosilicate synthesis [28] were reported to be efficient for the esterification of free fatty acids towards biofuel production.

Flint kaolin showed potential as a starting material for zeolite synthesis [43]. Recently, Oliveira et al. [24] demonstrated that it is possible to synthesize mesoporous aluminosilicate from this residue as an alternative source of silicon and aluminum, which enables it for various technological uses, including the synthesis of new catalytic materials, and which may lead to the development of specific integrated management projects. The synthesis of mesoporous aluminosilicates from low cost abundant natural clay minerals has been explored due to the similarities between their structural units and those of mesoporous materials [45,46].

As mentioned above, we have examples of work reporting the use of natural kaolinite as starting materials to synthesize Al–MCM-41. Generally, Al–MCM-41 exhibits a very high Si/Al molar ratio, and tends to show a decrease in the number of acid sites, thus reducing catalytic efficiency in reactions requiring a strong acid catalyst [24,28,38]. However, this problem can be solved by heteropoly acid impregnation [24,45] and anchoring of sulfonic groups [28] on the structure of Al–MCM-41 as an alternative method to obtain a high acid catalyst.

Sulfonic acids are organic compounds that have the functional group $-SO_3H$ attached to a hydrocarbon radical [28]. Mesoporous molecular sieves functionalized with this acid group have increased hydrophobicity (inherent in the presence of the alkyl chain) due to the organosulfonic group anchored on the surface of the material, making them very efficient organic-inorganic hybrid catalysts [28,47–49]. Mercaptopropyltrimethoxysilane (MPTS) is commonly used to functionalize mesoporous materials [28,47,50,51], by covalently bonding to the silane groups on the aluminosilicate surface, the sulfonic form ($-SO_3H$) being obtained only after an oxidation step of the $-SH$ precursor (mercapto group) by an oxidizing agent such as hydrogen peroxide [28,50].

Herein is an extension of our group's previous work, in which Lima and colleagues [28] synthesized Al-MCM-41 aluminosilicate from kaolinitic residue from the kaolin processing process and functionalized with sulfonic groups to obtain a very efficient catalyst for the conversion of free fatty acids into biodiesel. In another study, Oliveira et al. [24] synthesized a very active heterogeneous 10HPMo/AlSiM acid catalyst from kaolin *flint* for the reaction of eugenol acetylation with acetic anhydride. It was considered interesting to use a mesoporous aluminosilicate support synthesized from another available low cost kaolinitic tailings in the Amazon region, the *flint* kaolin [24]. In the present work, for the first time, we report the synthesis and characterization of mesoporous aluminosilicate (AlSiM) anchored MPTS from *flint* kaolin, as well as the use of this type of material as catalyst for the eugenol and acetic anhydride acetylation reaction for production eugenyl acetate, which can be used as a natural larvicide. Catalytic activity was evaluated for the esterification of eugenol and acetic anhydride as model substrates studying the effect of various reaction parameters, such as catalyst concentration, eugenol/acetic anhydride molar ratio, time and temperature, and the catalyst was regenerated up to five cycles. A kinetic study was also carried out and it was found that the acetylation reaction follows a first order reaction.

2. Results and Discussion

2.1. X-ray Fluorescence (XRF) Studies

Once the stoichiometric calculations of reagents required to form AlSiM is influenced by the percentage of silicon [52], it is fundamental to know about the chemical composition of the silica precursors. Table 1 shows the XRF (EDX; EDX-700, SHIMADZU, Kyoto, Japan) results. The amounts of SiO_2, Al_2O_3 and H_2O present in the *flint* kaolin (KF) are almost the same as the theoretical values obtained for kaolinite [21,25,43,44]. The values found for TiO_2 and Fe_2O_3 in KF, are mainly due to the greater presence of anatase (TiO_2) and hematite (Fe_2O_3). However, the high percentage of iron in KF may have occurred due to the isomorphic substitution of Al^{3+} by Fe^{3+} on the kaolinite octahedral sheet. [25,43,44]. Due to the dealuminization process caused by acid leaching of metakaolin *flint* (MF), in KF there is a mass a Si/Al ratio close to 1, in metakaolin *flint* leached (MFL) this ratio rises to 13 [28,38], while in the synthesized AlSiM it showed a Si/Al of 23 much higher than its precursor MFL.

Table 1. XRF analysis results of the raw material and samples prepared.

Samples	SiO$_2$	Al$_2$O$_3$	TiO$_2$	Fe$_2$O$_3$	SO$_3$	LF [a]	Si/Al
K [b]	46.54	39.50				13.96	1
KF	42.30	38.29	3.20	2.92		13.29	1
MF	49.64	37.92	2.50	2.78		1.25	1
MFL	73.98	4.74	3.71	1.62		15.89	13
AlSiM	86.72	3.21	3.52	1.10		5.23	23
(3)SO$_3$H/AlSiM	69.09	1.71	2.16	0.26	8.70	18.08	34
(5)SO$_3$H/AlSiM	61.39	1.55	1.83	0.10	11.06	24.07	34

[a] LF = loss on ignition; [b] K = theoretical kaolinite [21,25,43,44]; (3 or 5) SO$_3$H/AlSiM denotes (3 or 5 mmol) mercaptopropyltrimethoxysilane (MPT) functionalized in AlSiM.

The data in Table 1 also shows the presence of SO$_3$ for (3)SO$_3$H/AlSiM and (5)SO$_3$H/AlSiM, indicating that sulfonic groups are present in the AlSiM structure [28,50]

2.2. XRD Analysis

In the diffractograms (Bruker D8 Advance; Bruker Corp, Billerica, MA, EUA) presented in Figure 1a, For the KF sample we identify the main mineralogical components of the material where kaolinite (K) presents the highest intensity peak. This material also has low intensity components such as: quartz (Q), anatase (A) and hematite (H). These results are in agreement with the mineralogical compositions of the researched literature [25,43,44]. For the calcined sample (MF) at 750 °C for 2 h (Figure 1a), absence of characteristic peaks for kaolinite (close to 2θ = 12 e 24°). The lack of these characteristic peaks is due to the dehydroxylation of the Al(OH)$_6$ octahedral layer which becomes coordinated Al tetra and penta units during heat treatment, which makes this material more susceptible to acid leaching [21–23,25,44]. For leached metakaolin samples, no significant changes were observed in relation to their metakaolinitic *flint* precursor (MF).

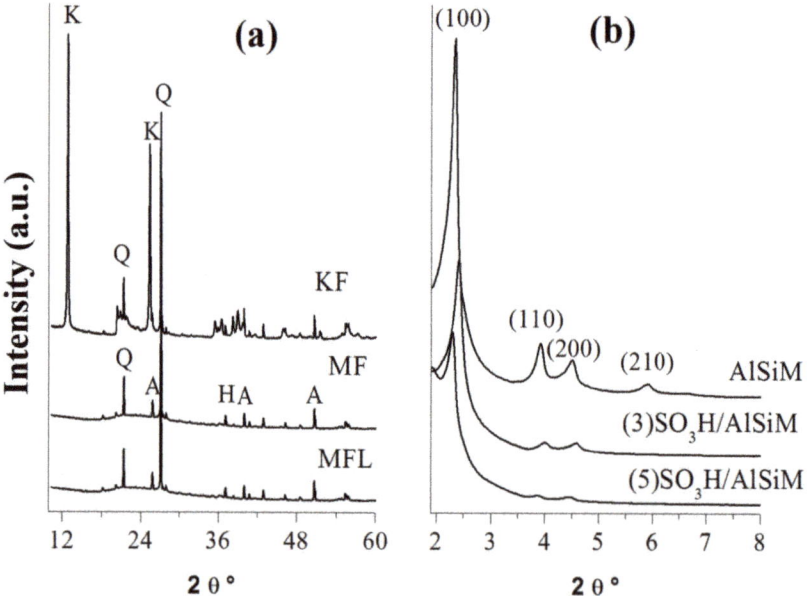

Figure 1. X-ray diffraction patterns: (**a**) *flint* kaolin (KF), metakaolin *flint* (MF), metakaolin *flint* leached (MFL), K = kaolinite, Q = quartz, A = anatase, H = hematite. and (**b**) aluminosilicates calcined (AlSiM) and functionalized with sulfonic group,(3)SO$_3$H/AlSiM and (5)SO$_3$H/AlSiM).

XRD patterns of AlSiM exhibited four peaks (1 intense and 3 weaks) at low 2θ values that can be indexed as (100), (110), (200) and (210) reflections characteristics of mesoporous materials with hexagonal arrangement of their cylindrical channels [40–42]; even after calcination (540 °C) there was no collapse of the structure. The Si/Al molar ratio of AlSiM obtained, which was close to 23, is suitable to form mesoporous silicas with well-defined structures, which have a high silicon content [28,38]. These results according with data found in literature, proving well-ordered formation of the structures with Si/Al molar ratio equal to 13.8 [53], Si/Al of 20 [28], Si/Al of 29.3 [42] and Si/Al of 32.1 [38].

For samples (3 or 5) SO$_3$H/AlSiM it is possible to observe characteristic peaks of the hexagonal system. As the AlSiM anchored MPTS content increased, it caused changes in the porous structure of the support as the four reflections widened and the peaks shifted to greater angles. This shift can be explained by a slight decrease in pore size resulting from the insertion of the sulfonic group into the linear pores of the AlSiM peak d_{100} becomes less intense and broad as reflections (110, 200 and 210) follow the same trend and disappear [28,48]. The absence of reflections at higher angles (>2θ) is an indication that the samples of (3 or 5)SO$_3$H/AlSiM have a structure with low structural order compared to AlSiM, however, there is a preserved hexagonal arrangement, which can be observed by the first d_{100} reflection plane even after functionalization with MPTS [28,48,50].

Table 2 shows the XRD data of the aluminosilicate samples studied from the reflection (100) obtained from each material, which correspond to the standards found in the literature [28,39–42,45,54]. The interplanar spacings (d_{100}) and the network parameters (a_0) calculated from the reflection (100) are also in good correspondence with literature data [39–42,45,54].

Table 2. Textural and structural characteristics.

Samples	Textural Property			Structural Property		
-	SSA (m^2 g^{-1}) [a]	V$_p$ (cm^3 g^{-1}) [b]	D$_p$ (nm) [c]	d_{100} (nm) [d]	a_0 (nm) [e]	W$_t$ (nm) [f]
KF	8.9	0.05	32.20	-	-	-
MF	9.78	0.05	31.11	-	-	-
MFL	433	0.56	1.20	-	-	-
AlSiM	1071	1.05	3.85	4.35	5.06	1.21
(3)SO$_3$H/AlSiM	998	0.78	3.25	4.07	4.70	1.45
(5)SO$_3$H/AlSiM	869	0.65	3.01	4.04	4.67	1.66

[a] SSA = specific surface area (BET method); [b] V$_p$ = pore volume (BJH method); [c] D$_P$ = pore diameter (BJH method); [d] d_{100} = interplanar spacing (100) (d_{100} = λCuKα/senθ) [28,34]; [e] a_0 = hexagonal unit cell parameter (2$d_{100}/\sqrt{3}$) [28,34]; [f] W$_t$ = mesoporous wall thickness (Wt = a_0 − D$_P$) [28,34].

2.3. Nitrogen Physisorption Experiments

The dehydroxylation of kaolinite by calcination led to the formation of metakaolin, considerably increasing SSA and Vp after acid leaching (Figure 2a and Table 1) (Micromeritics TriStar II model 3020 V1.03 apparatus (Micromeritics, Norcross, GA, USA). The MFL (433 m^2 g^{-1}) after the de-alumination process (Al$_2$O$_3$~88%, Table 1) had its SSA increased by approximately 43 times compared with its precursor (MF, 9.78 m^2·g^{-1}).

The isothermal forms of the four samples were completely different. MFL isotherm is a mixture of type II and IV according to Du and Yang [42], with an isotherm-like initial portion type I and a hysteresis loop mixture of H3 and H4, indicative of narrow slit pores. The preparation of porous material (MFL) from the MF by leaching involved the removal of part of the Al$_2$O$_3$ (Table 1) of MF by acid treatment, leaving pores within the structure of the MFL [40–42].

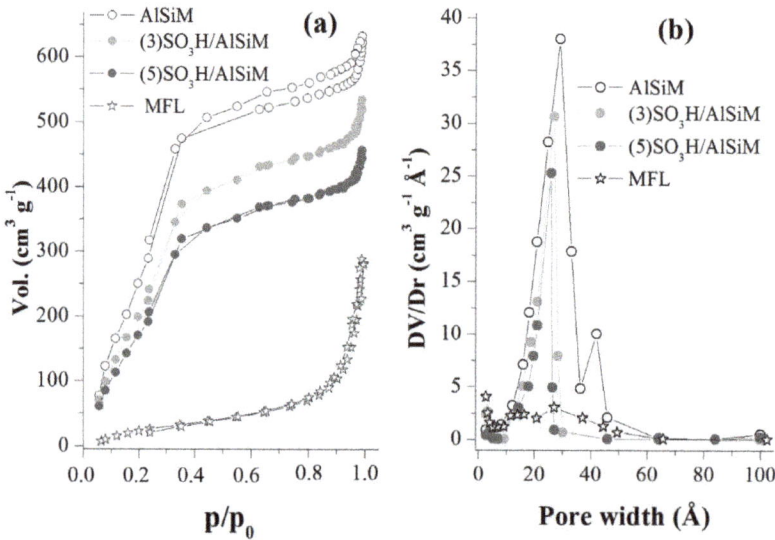

Figure 2. N_2 physisorption isotherms (**a**) and pore size distribution curves (**b**) of MFL, AlSiM and (3 or 5)SO$_3$H/AlSiM.

N_2 physisorption isotherms and pore size distribution of AlSiM and (3 or 5)SO$_3$H/AlSiM samples (Figure 2a,b) are all type IV according to the IUPAC type IV classification [55,56], an indicative of the formation of mesoporous material which structures were not compromised, even after functionalization, a behavior similar to those reported in the literature for materials of this type [28,33–35], with characteristic hysteresis cycle of uniform mesoporous material, corresponding to mesoporous materials obtained from metakaolin [39,41,42].

Figure 2a shows three distinct regions at low relative pressures. At $p/p_0 < 0.2$, N_2 adsorption occurs in monolayer and multilayer on the pore walls. Then, at average relative pressures, between $0.2 < p/p_0 > 0.3$, there is a strong increase in N_2 adsorption that is caused by capillary condensation of nitrogen inside the primary mesopores [39,42]. The formation of a hysteresis loop in p/p_0 equal to 0.3 is observed and results from an abrupt increase in the amount of adsorbed N_2, due to capillary condensation inside the mesoporous material [41]. With the filling of all primary mesopores by N_2, the third region starts in $p/p_0 > 0.3$, where the slope of the curve decreases tending to a plateau, which characterizes the adsorption in multilayers on the external surface of the solid [39,42]. Isotherms of mesoporous materials with a certain Al content, generally show a strong increase in the curve in p/p_0 >0.9, which is related capillary condensation in secondary mesopores due to the voids formed by crystalline aggregates in the hexagonal structure of the materials [39,42].

The pore size distribution shown in Figure 2b shows a narrow main peak centered at 2.8 nm, showing the presence of uniform mesopores. This result corroborates with that obtained by the XRD analysis (see Figure 1b), which revealed the formation of a highly ordered structure of the mesoporous molecular sieve (AlSiM) obtained in this work. The mesoporous material developed from KF as a source of Si and Al showed physical properties quite similar to the materials prepared using commercial metakaolin [39–42] or synthetic silica [22,33,34,57–60].

From the literature data [24,28,38–42] and the results of the textural analyzes obtained for the AlSiM synthesized in this work, it is possible to state that KF can be used as a low-cost silica source for the synthesis of mesoporous molecular sieves with uniform pores and with regular hexagonal arrangement.

Samples (3 or 5)SO$_3$H/AlSiM showed isothermal forms of N_2 physisorption analogous to AlSiM, with capillary condensation within the mesoporous close to the relative pressure range between $0.2 < p/p_0 > 0.3$. However, the adsorbed N_2 volume decreased with increasing MPTS content due to the structural disorder of the samples (Figure 1b). Consequently, SSA decreased from 1071 m² g⁻¹ (AlSiM) for 869 m² g⁻¹ (5)SO₃H/AlSiM), for the sample with the highest mmol of MPTS. Variation in MPTS content in the anchoring process did not significantly affect the average pore diameter but reduced the pore volume and increased the wall thickness of the samples (Table 2). The data of N$_2$ physisorption for these materials is in accordance with other mesoporous silicates functionalized with sulfonic groups [28,50,61]. Textural properties, such as SSA, V$_p$, and D$_p$, derived from N$_2$ physisorption measures are showed in Table 2. N$_2$ physisorption and XRD studies revealed that MPTS functionalization did not affect the mesoporous structures of these materials (Figure 1b,e and Figure 2a).

2.4. Fourier Transform Infrared Spectroscopy (FTIR) Experiments

AlSiM and (3 or 5)SO₃H/AlSiM FTIR spectra are shown in Figure 3. AlSiM FTIR (Shimadzu, Kyoto, Japan) spectrum show a broadband between 1228–1060 cm⁻¹ (asymmetrical stretching elongation vibrations O–Si–O). Also in the AlSiM spectrum, bands are observed at 967, 795 and 460 cm⁻¹, which correspond to the stretching vibration of silanols (Si–OH, tetracoordinated to silicon), angular deformation of the Si–O–Al bonds [39,42] and vibrational deformation of the tetrahedral bonds between Si–O–Si [48], respectively, after calcination of material these bands are also lightly shifted to higher frequencies [39,42].

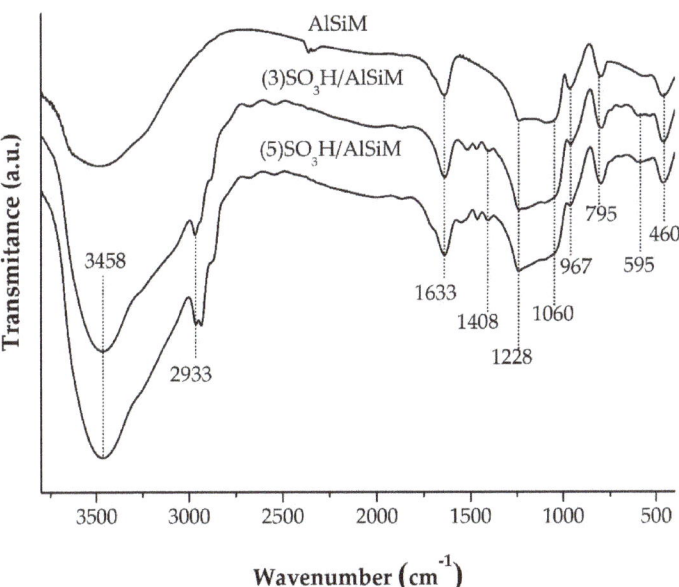

Figure 3. FTIR spectra of samples AlSiM e (3 or 5)SO₃H/AlSiM.

FTIR analysis shows variations in spectra caused by support functionalization by sulfonic groups. Figure 3 of all samples present bands around 3447 and 1633 cm⁻¹ that correspond to water adsorption by silanol groups located on the surface of the mesoporous material and the vibrational deformation of the adsorbed water, respectively [37–39]. The significant band at 2933 cm⁻¹ is induced by symmetrical and asymmetrical vibrations of –CH$_2$ present on the surface of the catalysts, showing that the MPTS reacted successfully with silanol groups and is covalently attached to the surface of the AlSiM [62]. The O–H strain vibration near 1633 cm⁻¹ can be seen in all samples [48]. The band near 595 cm⁻¹ may be due to C–S stretch vibration [61]. However, in the AlSiM sample, bands close to 2395 cm⁻¹ could be observed which could also correspond to the symmetrical and asymmetric stretching of the

C–O bond in the CO_2 molecule. These spectral results are quite consistent with those observed in the literature [61]. In the spectral range between 1300 e 1500 cm^{-1} of samples (3 or 5)SO_3H/AlSiM compared to the AlSiM matrix, absorption band around 1408 cm^{-1} is associated with the asymmetric and symmetrical stretching signals of O=S=O bonds of sulfonic groups [48,63]. The existence of the sulfonic acid group (SO_3H) in the samples (3 or 5)SO_3H/AlSiM was confirmed by the presence of an intense and wide absorption band at 3458 cm^{-1}, which corresponds to the vibration of the S-OH bond stretch, in addition to the 1408 cm-1 band attributed to asymmetric and symmetrical vibration connections (O=S=O), confirming the covalent anchoring of (SO_3H) to the surface of AlSiM [62,63] These results reveal that the sulfonic acid group was successfully anchored on the support surface, corroborating the data obtained by the XRF technique regarding the chemical composition (Table 1).

2.5. Thermal Analysis (TGA/DTG) Results

The existence of organosulfonic groups was monitored based on TGA/DTG (Shimadzu, Kyoto, Japan) analysis. The thermogravimetric analysis curves for AlSiM, (3 or 5)SO_3H/AlSiM are presented in Figure 4. Mass losses are more clearly identified by the DTG as it occurs on all TGA curves. The data obtained by TGA/DTG are summarized in Table 3, confirming mass losses in certain temperature ranges for both pure and functionalized AlSiM. In all samples an initial mass loss of 25 to 200 °C is observed, which many authors attributed to physically adsorbed water thermosorption on the surface of the materials [48,61,63].

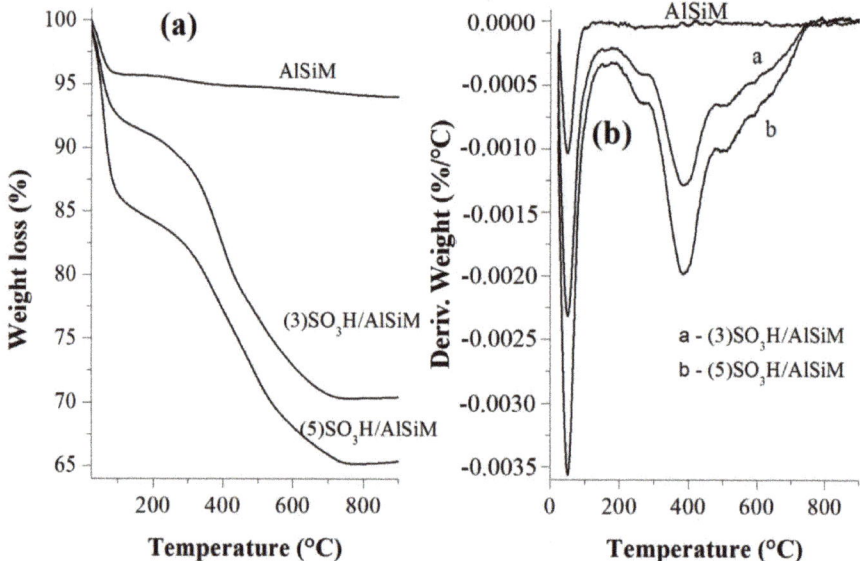

Figure 4. Thermal stability determined by TGA/DTG of AlSiM and (3 or 5)SO_3H/AlSiM (a) TGA and (b) DTG.

Table 3. Mass loss (%) with respective temperature ranges.

Samples	Mass Losses (%)			Total
	I (T < 200 °C)	II (200–430 °C)	III (T > 430 °C)	
AlSiM	3.96	0.99	0.80	5.23
(3)SO_3H/AlSiM	8.23	6.71	5.75	20.69
(5)SO_3H/AlSiM	12.65	8.09	6.55	27.29

The anchorage of organosulfonic groups in the AlSiM structure was followed by TGA analysis as shown in Figure 4. In Figure 4 we can see significant losses before 200 °C due to evaporation of water adsorbed some release of SO_2 from sulfonic acid groups (propyl-SH and propyl-SO_3, respectively) [49,61,64]. Two mass losses were observed in the range 200–470 °C, is attributed to the decomposition of organic matter, degradation of propyl groups present in the catalysts [49,61,64], towards higher temperatures is seen at the peak of organic matter elimination at 380 °C shown by the DTG curve. The final mass loss (between 500 to 900 °C) can be attributed to the higher thermal stability of the oxidized propyl sulfonic group [50] and the decomposition of residual organic groups and the condensation of neighboring silanol groups forming siloxane bridges, releasing water molecules [48,61,63]. The XRF analysis (Table 1) and FTIR (Figure 3) of functionalized material corroborate the results of the TGA, and the sample with the highest MTPS load was the sample with the highest mass loss, which is naturally caused by the larger amount of functionalized material.

2.6. Surface Acidity

The surface acidity of the AlSiM and functionalized (3 or 5)SO_3H/AlSiM solids were evaluated by acid–base titration [22,23,27]. Another way to evaluate the acidity of the materials was by means of FTIR methodologies and TGA/DTG technique after adsorption of pyridine as probe molecule on the material, which allows to characterize heterogeneous catalysts and to quantify acid sites efficiently [21,23,25,27].

Considering that the catalysts (3 or 5)SO_3H/AlSiM were the most active in the preliminary reaction studied, the number of acid sites obtained for these materials was due to the consumption of hydroxyls (titration) and adsorption of pyridine [21–23,25,27]. Table 4 shows the total hydroxyls and pyridines consumed, which indicates that all catalysts had more acid sites when compared to their support (AlSiM), because the number of acid sites (3 and 5 mmol) of MPT are present in the support structure (Table 1).

Table 4. Evaluation of acidity of the prepared materials and conversion of eugenol.

Amostras	(mmol H^+ g^{-1}) [a]	(μmol g^{-1}) [b]	Conv. (%) [c]
AlSiM	1.31		27
(3)SO_3H/AlSiM	5.93	295	97
(5)SO_3H/AlSiM	4.89	236	89

[a] Total surface acidity by titration [22,23]; [b] mole of pyridine by TGA/DTG analysis [25]; [c] conversion of eugenol (EugOH: AA molar ratio = 1:5, 2% catalyst, 80 °C, 30 min).

From Figure 5, TGA/DTG curves revealed that (3 or 5) SO_3H/AlSiM with adsorbed pyridine (Py) lost more mass than samples without adsorbed pyridine. In the range between 150 and 250 °C, the samples (3 or 5)SO_3H/AlSiM Py have losses that may be related to water and physically adsorbed pyridine, while there is a continuous loss that is attributed to chemically adsorbed pyridine above 250 °C. The methodology described by Nascimento et al. [21,25] Using TGA/DTG curves after pyridine adsorption was used to define the acidity and the number of acid sites of the catalysts.

In Figure 5, It was observed that the acidity of each material resulted directly from the number of mmol of MPTS functionalized in the support. Analyzing the Table 4, it was noted that the sample with lower MPTS content presented higher number of acid sites, (3)SO_3H/AlSiM (295 μmol g^{-1}) > (5)SO_3H/AlSiM (236 μmol g^{-1}), which may be justified by the greater accessibility of pyridine molecules to the acid sites of the functionalized material that had the highest textural properties (Table 2).

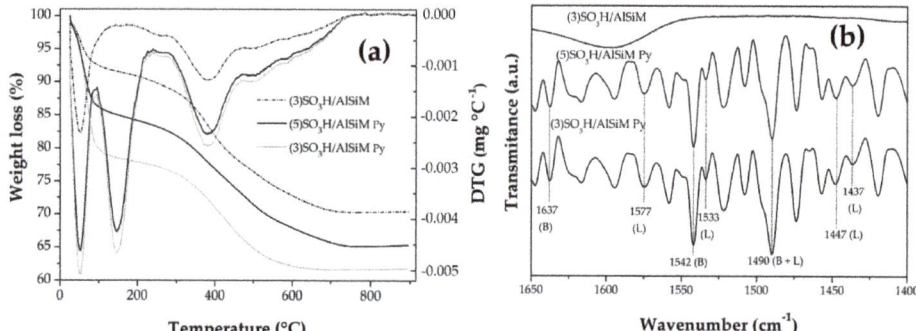

Figure 5. TGA/DTG curves (a) and FTIR spectra (b) of catalysts without/with adsorbed pyridine: (3) SO$_3$H/AlSiM and (3 or 5)SO$_3$H/AlSiM Py.

The concentrations of acidic sites measured by titration with NaOH were higher than those measured with pyridine. This may have been caused by the deposition of acidic sites (Al^{3+} or –SO$_3$H) on the pore surface of the material when the pyridine was adsorbed. Because pyridine is a bulky molecule, it may have caused a local blockage in the pore channels, making it difficult for other pyridine molecules to diffuse to the active sites present inside the solid. This difficulty in accessing pyridine to acidic sites significantly reduces the acidity values of the material. This relationship has already been reported in the literature in studies involving samples impregnated with pyridine and ammonia [65].

In FTIR spectra of the materials treated with pyridine (Figure 5b), bands related to pyridine physisorption were observed in 1637 and 1542 cm^{-1}, referring to the Brønsted acid sites; in 1577, 1533, 1447 and 1440 cm^{-1}, indicating the presence of Lewis acid sites; and in 1437 cm^{-1}, which refer to the acid sites of Brønsted and Lewis. In (3)SO$_3$H/AlSiM and (5)SO$_3$H/AlSiM samples, the generation of Lewis acid sites is possibly due to aluminum [23,25] present in the support (AlSiM) and Brønsted acid sites may be due to the presence of sulfonic acid group (SO$_3$H) [63], attested by the XRF technique (Table 1). Thus, the catalysts consist of Lewis acid sites and Brønsted acid site. This profile is in line with that presented by other mesoporous materials with sulfonic groups [63].

The results indicated the presence of substantial surface acidity in the catalysts and confirmed by FTIR spectra, and they clearly showed the presence of higher number of acid sites in (3) SO$_3$H/AlSiM due to higher SSA. Thus, the total acid sites are more dispersed by generating more accessible acid sites, compared to (5) SO$_3$H/AlSiM with 236 μmol g^{-1} acid sites distributed in smaller SSA (869 m^2 g^{-1}).

The acidity of the catalysts was evaluated in preliminary experiments on the eugenol acetylation reaction and these results are presented in Table 4. In the control tests, the conversion rate achieved over AlSiM was similar to the blank reaction (24% in 30 min). However, only the reactions catalyzed by the functionalized materials, (3) SO$_3$H/AlSiM > (5)SO$_3$H/AlSiM, showed significantly higher conversion rates than those obtained from the control experiments.

The observed trend (3) SO$_3$H/AlSiM > (5)SO$_3$H/AlSiM demonstrates the importance of catalytic acid site density and textural properties such as high SSA as well as V$_P$ and D$_P$ of the present catalyst pores play an important role in the diffusion of bulky molecules. such as eugenol and eugenol ester on the catalyst surface [10,66]. In the present case both catalysts have pore diameters of 3.25 nm ((3)SO$_3$H/AlSiM), 3.01 nm ((3)SO$_3$H/AlSiM), respectively, much larger than the maximum molecular size of the pore eugenol (~0.92 nm) and eugenyl acetate (~1.2 nm) (derived from Gaussian software as shown in Scheme 3). Therefore, there is sufficient space for diffusion of both reagents and products into the pores of both catalysts.

Low acid site concentration, together with lower SSA (869 m^2 g^{-1}) and V$_P$ (0.65 nm), are likely to be the most likely contributing factors to the low activity of (5)SO$_3$H/AlSiM. Although (5) SO$_3$H/AlSiM has more organic groups (–SO$_3$) in its structure confirmed by XRF (Table 1) and TGA (Figure 4) analyzes, it appears that the active sites were less accessible during both acidity study. as in catalytic testing (Table 4). However, the high catalytic activity of (3)SO$_3$H/AlSiM can be attributed to the presence of large SSA (998 m^2 g^{-1}) and V$_P$ (0.78 cm^3 g^{-1}), respectively, which contributed to both diffusion of these interacting large molecules. on the reagent-accessible active sites during the surface reaction. This facilitated the adsorption of polar substrates, such as eugenol and acetic anhydride, on the catalyst surface (Scheme 1) [10,67]. In addition, as eugenyl acetate is less hydrophilic than eugenol and acetic anhydride, (3)SO$_3$H/AlSiM showed a high tendency to adsorb these molecules and consequently showed higher product formation after prolonged reaction times [10,24,67].

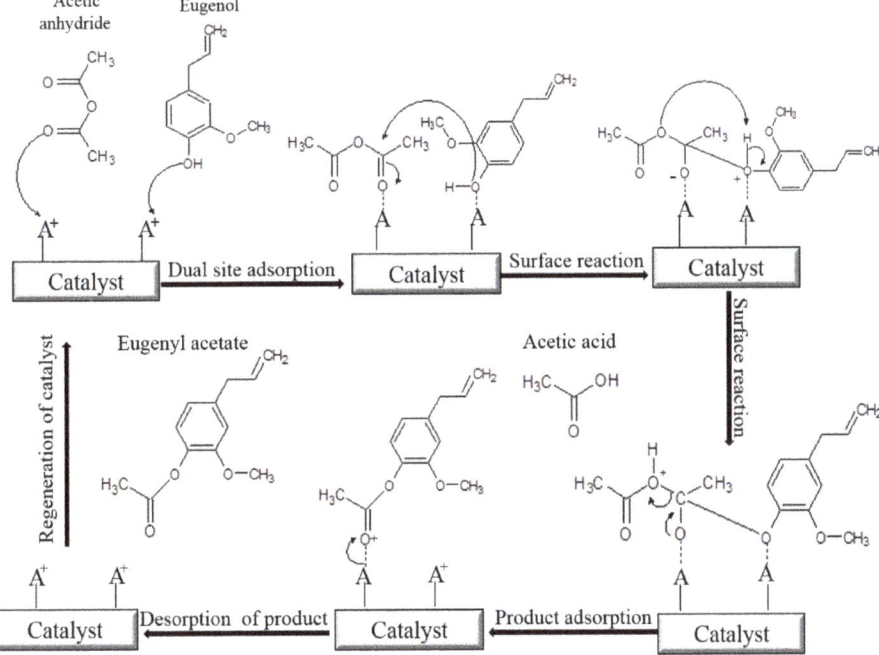

Scheme 1. Proposed mechanism for eugenol acetylation reaction on solid acid catalyst adaptation [10,24].

Overall, the results showed a good correlation with the effects of catalytic acid site density, textural properties and reagent molecule dimensions on catalyst reactivity, which are in good agreement with the trends found in the literature [10,17,31,32]. Effectively, (3)SO$_3$H/AlSiM has demonstrated activity that matches resins, *Amberlyst A-21* (95%) [32], Amberlite XAD-16 (98%) [31] and Lewatit® GF 101 (100%) [17]. As these results were satisfactory, (3) SO$_3$H/AlSiM was appointed for further investigation of various reaction parameters, such as the molar relationship between eugenol and acetic anhydride, catalyst amount, temperature and time and catalyst reuse to optimize reaction variables to achieve maximum eugenol conversion (confirmed by FTIR and GC-MS see supplementary material).

2.7. Eugenol Acetylation Reaction Mechanism with Acetic Anhydride

The eugenol (EugOH) (99% Aldrich, San Luis, Missouri, EUA) derivative ester was synthesized using acetic anhydride (AA) (Nuclear, São Paulo, SP–Brazil) in acid-solid catalyzed reactions following literature procedures [31,32]. Acetylation of eugenol with acetic anhydride is a limited equilibrium reaction. In order to overcome balance limitation, eugenol acetylation is usually performed by taking excess acetic anhydride to favor the direct reaction [5,67]. The mechanism is shown in Scheme 1.

A simple explanation for elucidating the pathway of the acetylation reaction mechanism on the acid catalyst surface [10,24] is shown in Scheme 1. Assuming the reaction takes place by the Langmuir–Hinshelwood–Hougen–Watson mechanism, where adsorbed reagents interact with acidic sites on the catalyst surface to form the products [10,24].

First, both reagents are adsorbed to the active sites (Brønsted and Lewis acids) of the catalyst [10,67]. In the next two steps, the surface reaction proceeds with the formation of the carbocation (electrophile) that is attacked by the nucleophile (eugenol) generating a protonated tetrahedral intermediate [68,69]. The transfer and rearrangement of protons (H^+) from hydroxyl (eugenol) to carbonyl oxygen (acetic anhydride) occurs in this intermediate which results in the formation of a rearranged tetrahedral intermediate. The subsequent stage takes place with the acetic acid leaving the tetrahedral intermediate, obtaining protonated eugenyl acetate [10,24]. Thus, the tetrahedral intermediate desorption (electron transfer from the catalyst acid site to the carbonyl oxygen, leading to bond disruption) and simultaneously the deprotonation of the ester and catalyst regeneration are initiated [10,70].

2.8. Effect of Temperature

The effect of reaction temperature was studied in the temperature range of 50 to 100 °C (Figure 6). A gradual increase in eugenol conversion from 85% to 99.3% was observed increasing the temperature from 50 to 80 °C after 40 min. As reported by Laroque et al. [31] and Santos et al. [67] at higher temperatures improves eugenol solubility and miscibility with anhydride, facilitating protonation of the acetic anhydride carbonyl group and eugenol nucleophilic attack on carbocation, resulting in higher eugenol conversion (Scheme 1). Compared to different reports, Lerin et al. [32], using *Amberlyst A-21* obtained 95% eugenol conversion at 95 °C. Another material known as zirconia dioxide (UDCaT-5) recorded 90% eugenol conversion at 110 °C Yadav [10]. As the boiling point of both eugenol and acetic anhydride is greater than 100 °C, and all reactions have been studied up to 100 °C, possibly there was no vapor loss during the liquid phase reaction. Hence, the temperature 80 °C was selected for acetylation for eugenol.

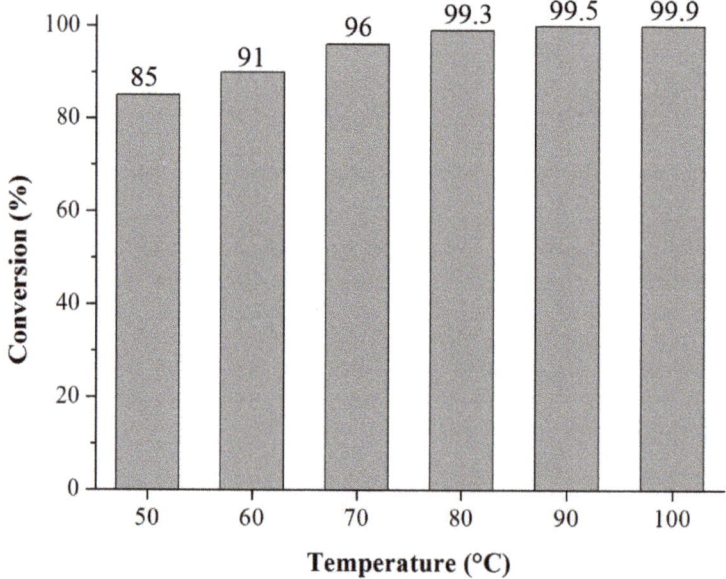

Figure 6. Percentage conversions of eugenyl acetate varying the reaction temperature. Reaction conditions: eugenol: acetic anhydride (1:5), reaction time 40 min and 2% catalyst.

2.9. Amount of Catalyst

Reactions without catalyst were performed, once the autocatalysis can occurs according to the mechanisms for the reaction studied [4,17], and the result for self-conversion was not higher than 24% (Figure 7). This reveals that the high activity of the catalyst in the reaction rate is directly proportional to the catalyst load based on the total reaction volume [10,17,31,71,72].

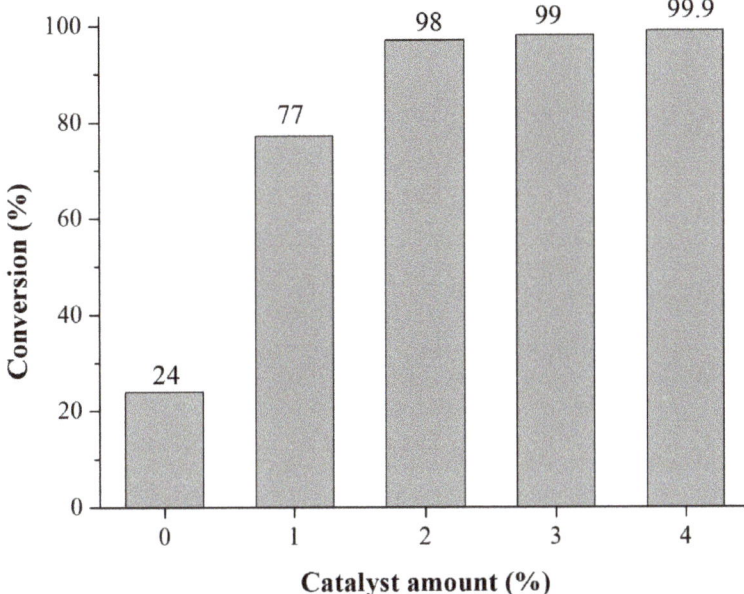

Figure 7. Percentage conversions of eugenyl acetate with different amount of catalyst. Reaction conditions: eugenol: acetic anhydride (1:5), reaction temperature 80 °C and reaction time 40 min.

Figure 7 shows the change in conversion (%) with the amount of catalyst charged during eugenol acetylation. The effect of the amount of catalyst was investigated by percent dosing from 1 to 4% mass (based on eugenol mass) of catalyst (3) $SO_3H/AlSiM$. Eugenol conversion has been observed to increase from 75% to its final value of 98% as catalyst dosage increases from 1% to 2%. However, this increase in eugenol conversion is expected due to the availability of more active sites that facilitate the acetylation reaction [10,31,32]. An additional increase in catalyst amounts resulted in not very appreciable changes in conversion, indicating that 2% catalyst represents the appropriate level of active sites for eugenol acetylation and that the equilibrium of the reaction is achieved.

2.10. Molar Ratio between Reagents

The reaction was studied with a molar ratio of eugenol to acetic anhydride from 1:2 to 1:7 over a 2% catalyst amount (relative to eugenol) at 80 °C for 40 min (Figure 8). The results showed a 61% conversion of eugenol with a 1:2 molar ratio of eugenol to acetic anhydride, which increased to 98% with the successive increase of the eugenol to acetic anhydride ratio to 1:5. However, by further increasing the eugenol to acetic anhydride ratio to 1:7, an almost similar conversion of eugenol (99%) was observed. Therefore, the eugenol to acetic anhydride ratio of 1:5 was selected for further studies.

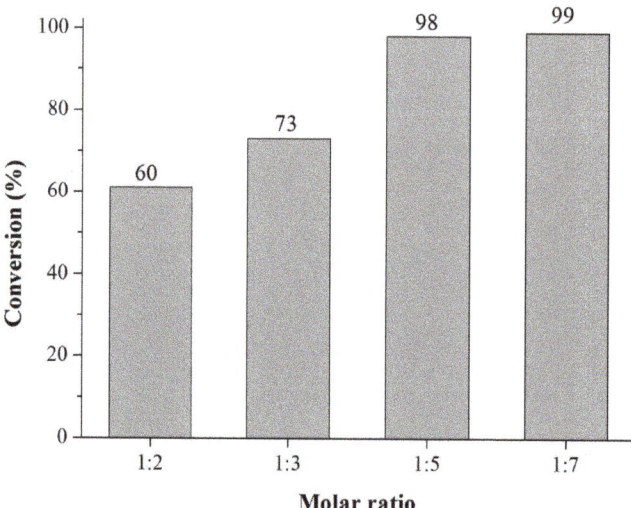

Figure 8. Percentage conversions of eugenyl acetate with different molar ratio of eugenol: acetic anhydride. Reaction conditions: reaction temperature 80 °C and reaction time 40 min and 2% catalyst.

2.11. Effect of Reaction Time

In order to evaluate the effect of reaction time on eugenol, reactions were allowed for varying durations of 10 to 50 min and the results obtained were shown in Figure 9. As the other reaction conditions were fixed, the reaction time was considered a dominant parameter controlling eugenol acetylation and therefore eugenol conversion. The eugenol conversion rate was 80% after 10 min of reaction but increases as the reaction duration lasts. The highest conversion rate (99.9%) was reached after 40 min. No further increase in eugenol conversion was observed after 50 min. Laroque et al. [31] investigated the acetylation of eugenol over molecular sieves 4Å and reported 90% conversion of eugenol in 2 h, a lower value than that obtained in this work.

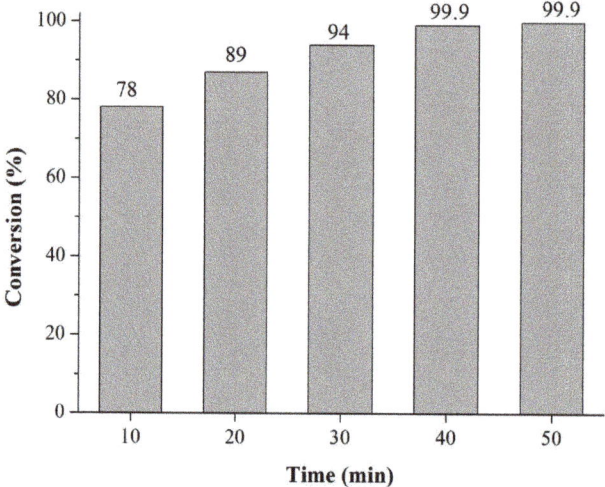

Figure 9. Percentage conversions of eugenyl acetate varying the reaction time. Reaction conditions: eugenol: acetic anhydride (1:5), reaction temperature 80 °C and 2% catalyst.

The optimum conditions for eugenol acetylation over (3) SO$_3$H/AlSiM were: eugenol molar ratio to acetic anhydride 1:5; temperature and reaction time 80 °C and 40 min respectively and amount of catalyst 2%.

From the high conversion values achieved with the catalyst synthesized here (99.9%) and based on the study of the kinetic parameters of the reaction, it is possible to conclude that the high conversion rates of eugenol, under the reaction conditions reported in this work (Figure 9), were higher than many others reported in literature (Table 5).

Table 5. A comparison of literature results for the conversion of Eugenol over various catalysts.

Catalyst	Solvent	T (°C)	R: M	T (min)	Conv. (%)	Reference
UDCaT-5	Toluene	110	1:5	240	90	[10]
Lipase (RML)	Chloroform	50	1:3	300	66	[11]
Lipozyme TL	Acetic anhydride	70	1:5	120	92.9	[5]
Molecular sieve 4Å	Acetic anhydride	60	1:3	120	90	[31]
Amberlyst A-21	Acetic anhydride	95	1:3	120	95	[32]
Lewatit® GF 101	Acetic anhydride	70	1:1	45	100	[17]
10HPMo/AlSiM	Acetic anhydride	80	1:5	40	99.9	[24]
(3)SO$_3$H/AlSiM	Acetic anhydride	80	1:5	40	99.9	Present work

Table 5 also presents some results with biocatalysts with maximum conversions of 66% [11] and 93% [5], even if using minimum molar ratio and mild reaction temperature, but reuse inefficiency and high process costs are the disadvantages of the biocatalyst [5,17].

The results revealed that the acid catalyst has a high potential in the acetylation reaction (99.9% conversion). These results are promising not only when compared to solid acid derived from kaolinitic wastes, but also to sulfonated metal oxides. For example, Yadav [10] prepared a zirconia-based catalyst, UDCaT-5, and evaluated it in the esterification of eugenol with benzoic acid, achieving high conversions, at the cost of reaction times and temperatures. Prolonged reaction times have also been observed for commercial catalysts such as molecular sieves 4Å [31] and *Amberlyst A-21* [32], as well as, full conversion was achieved with Lewatit® GF 101 [17] commercial catalyst. All of the mentioned catalysts used organic solvents, hexane and toluene, during recovery from reuse. Although the conversions obtained in these studies are high, even so, in most cases, the reaction times have always been prolonged when compared to the reaction times used in this work.

It is noteworthy that although hexane and toluene are ideal solvents for the extraction of organic substances, they have all the necessary and appropriate characteristics for catalyst washing. On the other hand, they are highly toxic and harmful to health. In this sense, it was decided to wash or regenerate the catalyst with anhydrous ethanol [24], as it has advantages such as lower cost and is not toxic to animal and human health.

Amazon *flint* kaolin, one of the by-products of the kaolin industry, which presents low cost and high availability, can be perfectly reused in the synthesis of mesoporous molecular sieves, minimizing the use of synthetic silicas, which are often toxic. In addition, kaolin *flint* can be used as a catalytic support or as an alternative heterogeneous catalyst in the acetylation of eugenol with routes to the production of eugenol acetate. Thus, the reuse of this by-product can generate a reduction in environmental impacts and costs in the synthesis of chemical substances with various applications.

2.12. Kinetic Studies

From the kinetic study (Figure 10a) of the different temperatures (50, 60, 70 and 80 °C) and times (10, 20, 30 and 40 min), an estimation was made for the activation energy. The conversion curves of eugenol as a function of time and temperature are observed in Figure 10a, there it is possible note the high performance of (3) SO$_3$H/AlSiM for eugenol acetylation when the temperature varies. Santos et al. [67] and Machado et al. [4] reported that increasing the temperature improves both the diffusion and the solubility of the reagents over the reaction time, leading to maximum conversion. As observed also in Figure 10a, high conversions (85%, 90%, 94% and 98%, respectively) were obtained in the first 20 min of reaction.

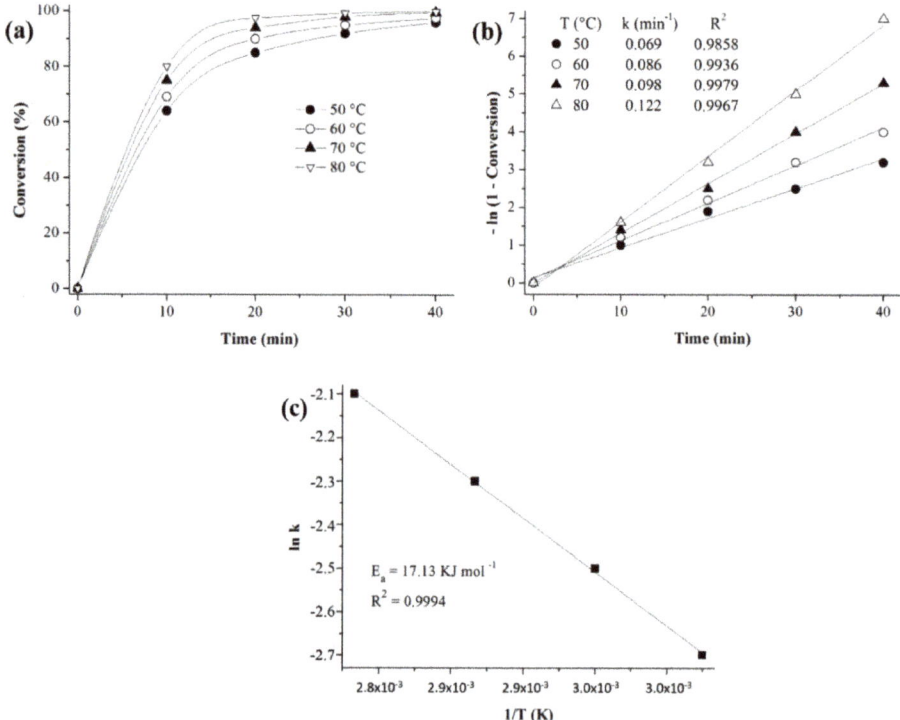

Figure 10. (a) Kinetic studies of the conversion of eugenol over the catalysts; (b) graphs of −ln (1−conversion) versus time at different temperatures for calculations of velocity constants and reaction order analysis; (c) Arrhenius graph for calculating the activation energy for eugenol esterification reaction at different temperatures on the catalyst.

In this work, we determine the order of the esterification reaction of eugenol with acetic anhydride, using the classical definitions of chemical kinetics and considering eugenol as a limiting reagent [10,31] To determine the order of reaction by the elementary kinetic theory, we proposed a complete conversion of eugenol. According to data from Figure 10a, it can be concluded that there is a first order dependence between reaction rate and eugenol concentration, because the fit shows a linear relationship between all experimental data when −ln (1−conversion) is plotted against reaction time. According to the standard deviations (R^2) of 0.9858, 0.9936, 0.9979 and 0.9967 for acetylation carried out at 50, 60, 70 and 80 °C, respectively (Figure 10b). It is observed that the kinetic data analyzed are consistent, as it is a first order reaction in relation to eugenol [10,31]. From the values of k shown in Figure 10b, plotted ln (k) versus 1/temperature (K) (Figure 10c) and the apparent activation energy (E_a) was calculated, which was 17.13 kJ mol^{-1} with good linear regression (R^2 = 0.9994).

The calculation of activation energy can be found in some studies on esterification of eugenol in heterogeneous reaction. For example, Yadav [10] tested the UDCaT-5 catalyst in a heterogeneous esterification reaction and reported a first order reaction in relation to eugenol and an activation energy equivalent to 39.13 kJ mol^{-1}. Laroque et al. [31] used two solid catalysts, molecular sieve 4 Å and Amberlite XAD-16, and the kinetic study of acetylation showed a first order reaction with activation energies of 10.03 kJ mol^{-1} and 7.23 kJ mol^{-1}, respectively. Oliveira et al. [24] applied the

10HPMo/AlSiM catalyst in the same reaction in the acetylation kinetics study found a first order reaction and activation energy equal to 19.96 kJ mol^{-1}. Thus, the activation energy obtained in the present work is in the same order of magnitude as those reported in previous works.

The catalyst synthesized here exhibited intermediate activation energy in relation to the different types of solid acid catalysts, related to the esterification of eugenol, which were mentioned in the present manuscript. However, with regards to the reaction conditions, the proposed system operated at a relatively moderate temperature of 80 °C, while those reported had reached 110 °C [10]. In addition, the results of the catalytic tests show that high conversions were achieved by our catalyst within 40 min, and that others from literature achieved comparable conversion only within 4 h reaction. Obviously, the catalytic performance of (3)SO$_3$H/AlSiM is comparatively superior to that of the catalysts we listed for analysis in our study. thus, it is possible to state that (3)SO$_3$H/AlSiM is an excellent catalyst for the production of eugenyl acetate, even with an activation energy of 17.13 kJ mol^{-1}.

2.13. Catalyst Reuse

The use of heterogeneous catalysts is motivated by several factors, including the possibility of reuse in successive cycles, making them essential for reducing production costs and environmentally friendly [10,17,31,32]. Based on this context, the prepared catalyst reuse cycles were evaluated under the same conditions used in the kinetic study. After each cycle of reaction, the catalytic solid was recovered at 6000 rpm for 10 min of centrifugation, washed with ethyl alcohol to remove any organic compounds eventually retained on the support surface, dried in an oven at 150 °C for 2 h subjected to the next catalytic cycle. Ethanol was chosen to perform the washing step based on its good performance, as previously demonstrated for 10HPMo/AlSiM catalyst [24]. The conversion of eugenol after each run (Figure 11) gradually decreases from 99.96% to 92% at the end of the fifth run, revealing an overall reduction close to 8%. These results are slightly better those reported by Oliveira et al. [24] using a solid acid catalyst prepared by anchoring HPMo on an AlSiM mesoporous support to convert eugenol under consecutive runs, revealing a reduction in catalytic activity of approximately 10% after the fifth recycling (90%), compared to the initial value of 99.9%. These eugenol conversion results are excellent, especially considering the low cost associated with this catalyst and the good conversions achieved up to the fifth cycle. These results are even superior to the reaction without the presence of a catalyst (Figure 7), indicating that material can be reused for more reuse cycles. Therefore, it is a clear indication that the catalyst is acting on a heterogeneous route with good catalytic activity using the simple ethanol wash procedure.

2.14. Characterization of Reused Catalyst

After the fifth catalyst reuse cycle, its integrity was assessed by DRX and FTIR techniques. To determine the structural stability of the mesoporous catalyst under eugenol acetylation reaction conditions, the XRD patterns of the samples before ((3) SO$_3$H/AlSiM) and after reuse ((3)SO$_3$H/AlSiM R) were compared; these results are shown in Figure 12a. After the fifth reuse cycle, a slight decrease in the reflection intensity (100) and a partial disappearance of the reflections (110) and (200) of the reused catalyst can be noted possibly indicating the collapse of the structure to the catalyst used, even so the hexagonal symmetry of the material is still preserved [22,60]. The X-ray diffraction results of (3)SO$_3$H/AlSiM R (Figure 12a) shows little similarity to the diffractogram pattern of the new catalyst. The three reflection peaks of the plane (100) characteristic of the hexagonal structure are observed; as well as two other less intense peaks inherent to the planes (110) and (200) reflections, indicating a material pore system poorly organized [22,60]. This behavior could be associated to molecules adsorbed on the surface as well as inside the channels, consequently indicating a structural disorder of the support (Figure 12a), being one of the reasons for catalyst deactivation during reaction [27]. This result is favorable once, even after successive reuse cycles and despite the recognized fragility of the material, it is observed that its hexagonal structure is preserved [60].

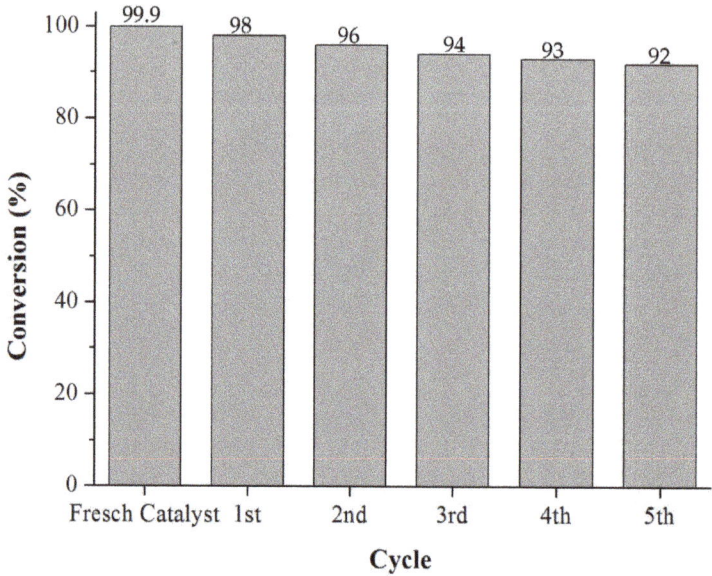

Figure 11. Reaction conditions: eugenol: acetic anhydride (1:5), reaction temperature 80 °C, reaction time 40 min and 2% catalyst.

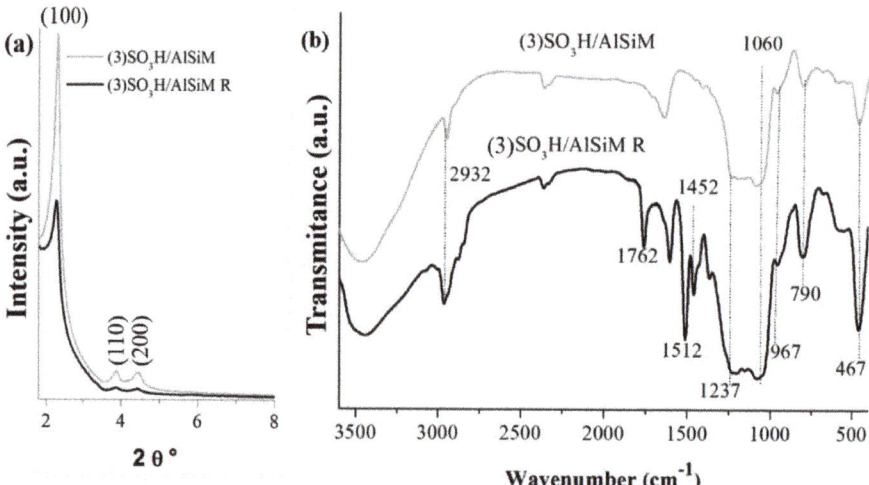

Figure 12. (a) Comparison of new catalyst XRD standards, (3)SO$_3$H/AlSiM and after reuse, (3)SO$_3$H/AlSiM R; (b) Comparison of new catalyst FTIR spectra, (3)SO$_3$H/AlSiM, and after reuse (3)SO$_3$H/AlSiM R.

The mechanism of adsorption of molecules on the catalyst surface as well as the numerous stages of simultaneous surface reactions was well understood in Scheme 1 and are one of the possible factors (operating conditions) that favor the catalyst deactivation process. The presence of prominent bands of organic groups is clearly noted from the FTIR result. The FTIR Spectrum (Figure 12b) of (3)SO$_3$H/AlSiM R showed prominent bands in 2932, 1765, 1513 cm^{-1} which were attributed to the stretching vibration of CH, CH$_2$/CH$_3$, C=O and presence of aromatic ring (C=C) [4,73–76] demonstrating the binding of the esterification product to the surface of the reused catalyst. However, these characterizations showed that there were many differences between new and used catalysts, partially indicating that the catalyst exhibited good structural stability and activity for the eugenol reaction process.

3. Materials and Methods

3.1. Materials

All chemicals used in this experiment were analytical grade and used without further Darmstadt, Germany purification: eugenol (EugOH) (99%, Sigma-Aldrich, Aldrich, San Luis, Missouri, EUA) and acetic anhydride (AA) (nuclear, São Paulo, SP–Brazil), sulfuric acid (H_2SO_4) (98%, ISOFAR, Duque de Caxias, Rio de Janeiro, Brazil) 3-mercaptopropyltrimethoxysilane (MPTS) (synthetic grade, Sigma-Aldrich, Aldrich, San Luis, Missouri, EUA), cetyltrimethylammonium bromide (CTABr, Sigma-Aldrich, Aldrich, San Luis, Missouri, EUA), anhydrous ethanol (EtOH) (98%) (synthetic grade, nuclear, São Paulo, SP–Brazil) and sodium hydroxide (NaOH), dichloromethane (CH_2Cl_2), diethyl ether (Et_2O) peroxide hydrogen (H_2O_2), toluene (all synthetic grade, VETEC, Darmstadt, Germany). The *flint* kaolin from the Capim River region (Pará-Brazil), kindly supplied by a partner of Institute of Geosciences (UFPA), was crushed and the sandy fraction separated by sieve retention. The fraction smaller than 62 µm was then collected, diluted in distilled water and centrifuged to separate the silt fraction, obtaining the clay fraction based on the previous work by Nascimento et al. [21,25].

3.2. Preparation of Catalysts

3.2.1. Thermal and Acid Treatments of Amazon *Flint* Kaolin

Due to its high content of octahedral aluminum, kaolin is resistant to acid leaching. The calcination of kaolin leads to the formation of the metakaolin phase which makes this material more susceptible to leaching of aluminum and iron cations from the octahedral layer [21,25]. Thus, the clay fraction of the *flint* kaolin (KF) was calcined at 750 °C for 5 h resulting in metakaolin *flint* (MF). The MF sample was leached in a solution of H_2SO_4 (2.5 mol L^{-1}) in a proportion of 1:10 (MF: H_2SO_4) for 1 h at 90 °C and then washed with 100 ml of H_2SO_4 (0.5 mol L^{-1}) and water until pH = 7 is reached and then dried at 120 °C for 12 h. The metakaolin sample obtained was designated as MFL (Scheme 2) [24,28].

3.2.2. Synthesis of Mesoporous Aluminosilicate (AlSiM)

The synthesis of AlSiM was carried out according to the methodology recommended in the literature that obtains mesoporous material at 110 °C in 24 h by hydrothermal treatment [24,28]. The procedure used consists of 3 g of MFL, 0.6172 g of sodium hydroxide, 1.8168 g of hexadecyltrimethylammonium bromide and 134.7 ml of distilled water. The resulting mixture remained stirred for another 24 h and, subsequently, was placed in stainless steel autoclave-coated teflon vessel. The set was subjected to hydrothermal treatment (110 °C for 24 h). The material obtained was filtered and washed with distilled water and dried in a muffle at 100 °C for 4 h and calcined at 540 °C for 5 h, with a heating ramp of 5 °C min^{-1}. The mesoporous material resulting from KF as a source of silica was called AlSiM. A summary of the procedure for preparing AlSiM from natural KF is shown in Scheme 2 [24,36,37,42].

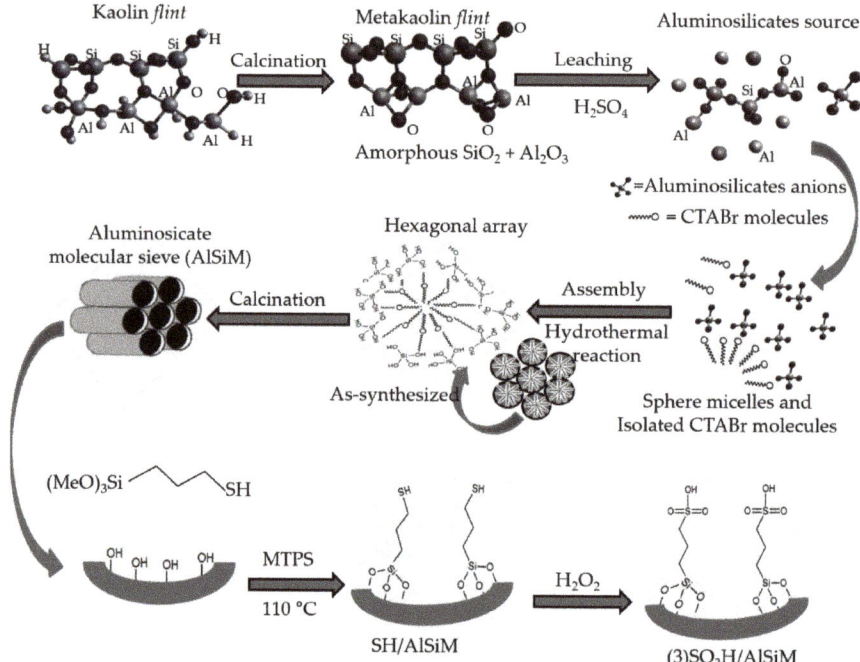

Scheme 2. Possible formation of AlSiM from natural *flint* kaolin followed by functionalization with MPTS.

3.2.3. Functionalization of AlSiM with MTPS

The AlSiM functionalization process was based on the procedure described by Lima et al. [28] (Scheme 2). Initially, AlSiM was dried at 110 °C for 24 h; then 1 g AlSiM was dispersed in a functionalizing solution with 3 and 5 mmol MPTS in 30 mL of toluene while stirring for 12 h at 110 °C under reflux. After completion of this process, the material was filtered, washed with 50 mL of toluene and dried at 70 °C for 12 h. Excess MPTS, not anchored in AlSiM, was extracted on a Soxhlet extractor with a mixture of CH_2Cl_2/Et_2O (50% v/v) for 12 h, purified solid material (SH/AlSiM) was filtered and dried. Anchored –SH groups were oxidized by immersing 1 g SH/AlSiM in 17 ml H_2O_2 for 12 h at room temperature. The oxidized material was then washed with H_2O/EtOH (50% v/v) and extensively with distilled water to neutral pH, dried at 60 °C and stored in a desiccator for further characterization. After all this process, the catalyst was denoted as (x)SO_3H/AlSiM, where x = mmol of MPTS.

3.3. Characterization

The chemical compositions of the samples were obtained with Shimadzu EDX-700 energy dispersive X-ray spectrometer (EDX; EDX-700, SHIMADZU, Kyoto, Japan), with a rhodium X-ray source tube (40 kV, SHIMADZU, Kyoto, Japan). For each analysis, approximately 500 mg (powder) of each sample was deposited in a lower sample holder made of polyethylene film in order to determine the silica content present in the precursor kaolin *flint*.

X-ray diffraction analysis were performed on a Bruker D8 Advance diffractometer (Bruker D8 Advance; Bruker Corp, Billerica, MA, EUA), using powder method, at a 1° < 2θ >10° interval. Cu Kα (λ = 1.5406 Å, 40 kV e 40 mA) radiation was used. The 2 θ scanning speed was 0.02° min^{-1}. The equation $a_0 = 2d_{100}/\sqrt{3}$ was used to calculate the distance (a_0) between pore centers of the hexagonal structure [28,34].

N$_2$ adsorption–desorption isotherms were obtained at liquid nitrogen temperature using a Micromeritics TriStar II model 3020 V1.03 apparatus (Norcross, GA, USA). Before each measurement, the samples were outgassed at 200 °C for 2 h. The specific surface area (SSA) was determined according to the standard Brunauer–Emmett–Teller (BET) method. Pore diameter (D_p) and pore volume (V_p) were obtained by the Barret–Joyner–Halenda (BJH) method [55,56,77]. The pore wall thickness (W_t) was calculated according to equation $W_t = a_0 - D_p$ [28,34].

The spectroscopy analysis in the infrared region was performed in a Fourier Transform Infrared (FTIR) spectrophotometer from Shimadzu (Kyoto, Japan), model IRPrestige-21 A, using KBr pellets. These pellets were prepared by mixing 0.2 mg of sample with a sufficient amount of KBr to achieve a concentration of 1% by mass. The spectra were obtained in the spectral region from 4000 to 400 cm^{-1} with resolution of 4 and 32 scans.

The FTIR technique was used to verify the nature of the acidic sites (Brønsted and Lewis) on the surface of the catalyst having pyridine adsorbed as a probe molecule [24,27,78,79]. The identification of acid sites was performed by preheating about 50 mg of the sample at 120 °C for 90 min, before pyridine treatment with probe molecule. The loosely filled sample was brought in contact with pyridine (about 0.1 cm^3) directly. Then the sample was kept in a hot air oven at 120 °C for 1 h to remove physiosorbed pyridine. After cooling the catalyst sample, the FTIR spectrum was recorded in the spectral range 1700 and 1400 cm^{-1} with 256 scans and at a resolution of 4 cm^{-1} using KBr background [24,27,78,79].

The thermal decomposition of the AlSiM and (3 or 5) SO$_3$H/AlSiM samples was performed on a Shimadzu DTG-60H (Kyoto, Japan) thermogravimetric (TGA/DTG) analyzer. About 10 mg (powder) of each sample is used for each analysis and placed in a platinum pan and heated from 25 °C to 900 °C, with a heating rate of 10 °C min^{-1} and inert gas flow of 50 mL min^{-1}.

TG Analyses also were used to quantify the number of acid sites. Specifically, the samples prepared for evaluation of the presence of acidic sites in the infrared were submitted to thermal analysis and the TGA/DTG curves of the samples without adsorption and with adsorbed pyridine (Py) were acquired. The difference in mass of the sample before and after being submitted to the adsorption of pyridine was used to calculate the number of acid sites. The value of this difference corresponds to the mass of pyridine adsorbed where each mole of pyridine equals one mole of the acid site present on the surface of the catalyst. From these data the number of mmol of pyridine (nPy) per gram of sample was mathematically determined according to the method proposed by Nascimento et al. [21,25].

Acid-base titrations were used to determine the surface acidity of the catalyst [22,23,27]. In a typical measurement, 0.1 g of solid was suspended in 50 mL of 0.1 M NaOH. The suspension was stirred for 24 h at room temperature and titrated with 0.1 M HCl in the presence of phenolphthalein. The surface acidity of the catalyst was expressed in mmol H$^+$ g^{-1} of catalyst.

3.4. Catalytic Tests

Acetylation of Eugenol

The acetylation reaction of eugenol using the catalyst (3) SO$_3$H/AlSiM was carried out in a PARR 4871 multi-reactor (Parr Instrument Company, Moline, IL, USA). All experiments were carried out under different reaction temperatures (50, 60, 70 and 80 °C), molar proportions eugenol/acetic anhydride (1:5), quantity of catalyst (2% w/w in eugenol), at 500 rpm for a maximum period of 40 min (Scheme 3). The (3) SO$_3$H/AlSiM catalyst was previously dried at 130 °C for 2 h before the reaction. After each experimental run, the catalyst was separated from the reaction medium by centrifugation (6000 rpm for 10 min) and the reaction conversion was determined by gas chromatography (GC/MS).

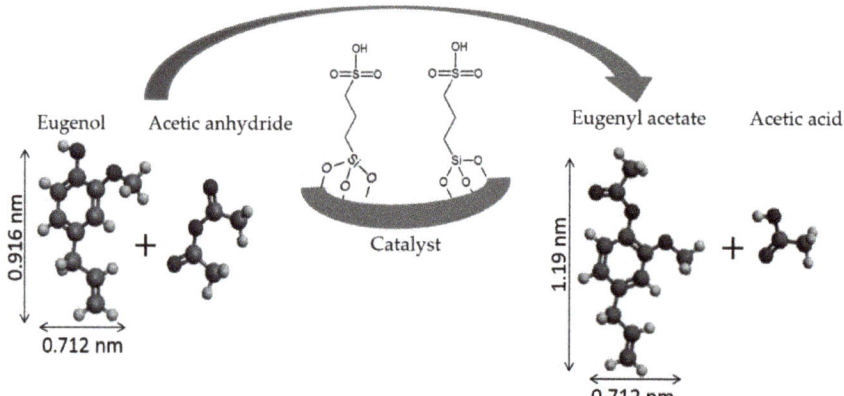

Scheme 3. Design of acetylation of eugenol with acetic anhydride on the catalyst during catalytic tests, molecular size of eugenol (l = ~0.916 nm, w = ~0.712 nm) and eugenyl acetate (l = ~1.19 nm, w = ~0.712 nm) by Gaussian software.

The reaction products were quantified by gas chromatography (GC/MS), using Shimadzu equipment (Shimadzu QP2010 plus instrument, Shimadzu Corporation, Kyoto, Japan). A capillary column of fused silica Rtx-5MS (30 m × 0.25 mm × 0.25 µm) was used. Helium (He) was used as carrier gas in a flow of 1.2 mL min^{-1}. The injector and detector temperatures were 200 °C and 240 °C, respectively. 1.0 µL of solution of products in 1.5 mL of dichloromethane were injected in the flow division mode (20:1). The programming of the column temperature for the analyzes started at 60 °C, followed by heating at the rate of 3 °C min^{-1} until reaching 240 °C. The mode of operation of the EIMS (electron ionization mass spectrum) was with energy of 70 eV, temperature of the ion source and connection parts, 200 °C. The chemical constituents of the reaction were identified by comparison with the substance library (NIST-11, FFNSC-2) [80] and with data from the literature [73,74]. This analytical procedure was performed at Adolpho Ducke Laboratory—Museu Paraense Emílio Goeldi.

The spectra were scanned within the range m/z 40–300. GC/MS analysis shows the presence of the ester at a retention time of 31.6 min (Figure S2 and Table S1). The mass spectrum of ester product showed a molecular ion peak at m/z 206 that corresponded to molecular formula of eugenyl acetate ($C_{14}H_{14}O_3$). Cleavage of the ester gave the fragment peak at m/z 164 related to the eugenol part [M+H]$^+$. Other fragments ions were also observed at m/z 41, 57, 91, 103, 131 and 149 [4,73–76] (Figure S3 and Table S1, see Supplementary Material).

Samples of product (eugenyl acetate) was subjected to FTIR analysis and the spectra were obtained using a Shimadzu model IRPrestige-21A Spectrophotometer. For each sample, the mean of 32 scans in the range 400–4000 cm^{-1} and the resolution of 4 cm^{-1} was done. The measurements of the samples were normalized by the air background.

The wavelength values obtained in the analyses of eugenyl acetate they are represented in Figure S1, where is observed the presence of the characteristic carbonyl band of the ester linked to the aromatic ring at 1765 cm^{-1}, confirming that only the acyl group was added to the eugenol molecule [4,73,76].

4. Conclusions

Based on the characterization techniques employed to study the properties of the new catalyst synthesized by this work, it can be concluded that *flint* kaolin was successfully used as an alternative source of silicon and aluminum for mesoporous catalytic support (AlSiM) synthesis, once that the material obtained has intrinsic properties of mesopore material. The anchoring of sulfonic groups in the AlSiM structure enabled (3) SO$_3$H/AlSiM to be active for eugenol acetylation with a maximum conversion of 99.9%. The performance of the catalyst was attributed to its high SSA (998 m^2 g^{-1}), Vp (0.78 cm^3 g^{-1}), Dp (3.25 nm) and strong surface acidity (295 μmol g^{-1} of pyridine). material, which favors diffusion regarding the conversion of the substrate and the product. Although successive reuse of the catalyst showed a slight decrease in activity, it nonetheless showed 90% conversion recovery over the fourth cycle, i.e., 92% of the original value (99.9%). The decrease in catalytic activity after each catalyst reuse cycle results from the adsorption of molecules from the reagent or product to their active sites. The integrity of the catalyst before and after the fifth cycle of reuse was confirmed by XRD, which results confirmed that the mesoporous structure was preserved. The good stability, the recyclability and the high catalytic activity in converting eugenol were remarkable characteristics of this new catalyst, which were obtained by a simple method with reduced reaction time and cleaner and lower cost processes. Therefore, that new eco-friendly heterogeneous catalyst was efficient to synthesize a molecule with potential larvicidal activity.

Supplementary Materials: The following are available online at http://www.mdpi.com/2073-4344/10/5/478/s1. Figure S1: Eugenol and Eugenyl acetate FTIR spectra, Figure S2: Mass spectrometry (GC/MS) (a) eugenol, (b) eugenol and eugenol acetate, (c) acetate eugenol, Figure S3: Chromatograms (a) eugenol, (b) eugenol and eugenol acetate (autocatalysis), (c) eugenol and eugenol acetate (catalyzed with (3)SO3H/AlSiM), (d) eugenol acetate (catalyzed with (3)SO3H/AlSiM). Table S1: Chemical characterization of eugenol and eugenyl acetate by GC/MS.

Author Contributions: Conceptualization, L.A.S.d.N.; methodology, E.H.d.A.A., C.E.F.d.C. and G.N.d.R.F.; formal analysis, A.d.N.d.O., L.H.d.O.P. and E.T.L.L.; investigation, A.d.N.d.O., E.T.L.L. and L.H.d.O.P.; resources, E.H.d.A.A., J.R.Z. and C.E.F.d.C.; data curation, R.L. and L.A.S.d.N.; writing—original draft preparation, A.d.N.d.O. and E.T.L.L.; writing—review and editing, J.R.Z. and R.L.; visualization, R.L., G.N.d.R.F. and L.A.S.d.N.; supervision, R.L., C.E.F.d.C., G.N.d.R.F. and L.A.S.d.N.; project administration, R.L. and L.A.S.d.N.; funding acquisition, G.N.d.R.F. and L.A.S.d.N. All authors have read and agreed to the published version of the manuscript.

Funding: This research was funded by Banco da Amazônia grant number 2018/212 and CNPQ, grant number 432221/2018-2.

Acknowledgments: The authors would like to thank the laboratories that supported this work: Laboratory of Research and Analysis of Fuels (LAPAC/UFPA), Laboratory of Catalysis and Oil Chemistry (LCO/UFPA) and LABNANO-AMAZON/UFPA. CAPES/UNIFAP and PROPESP/UFPA for the financial support.

Conflicts of Interest: The authors declare no conflict of interest.

References

1. Smith, L.B.; Kasai, S.; Scott, J.G. Pyrethroid resistance in Aedes aegypti and Aedes albopictus: Important mosquito vectors of human diseases. *Pestic. Biochem. Physiol.* **2016**, *133*, 1–12. [CrossRef] [PubMed]
2. Barbosa, J.D.F.; Silva, V.B.; Alves, B.; Gumina, G.; Santos, R.L.C.; Sousa, D.P.; Cavalcanti, S.C.H. Structure—Activity relationships of eugenol derivatives against Aedes aegypti (Diptera: Culicidae) larvae. *Pest Manag. Sci.* **2012**, *68*, 1478–1483. [CrossRef] [PubMed]
3. Govindarajan, M.; Rajeswary, M.; Benelli, G. Chemical composition, toxicity and non-target effects of Pinus kesiya essential oil: An eco-friendly and novel larvicide against malaria, dengue and lymphatic fi lariasis mosquito vectors. *Ecotoxicol. Environ. Saf.* **2016**, *129*, 85–90. [CrossRef] [PubMed]
4. Machado, J.R.; Pereira, G.N.; de Oliveira, P.D.S.; Zenevicz, M.C.; Lerin, L.; de Oliveira, R.D.R.B.; de Holanda Cavalcanti, S.C.; Ninow, J.L.; de Oliveira, D. Synthesis of eugenyl acetate by immobilized lipase in a packed bed reactor and evaluation of its larvicidal activity. *Process Biochem.* **2017**, *58*, 114–119. [CrossRef]
5. Silva, M.J.A.; Loss, R.A.; Laroque, D.A.; Lerin, L.A.; Pereira, G.N.; Thon, É.; Oliveira, J.V.; Ninow, J.L.; Hense, H.; Oliveira, D. Lipozyme TL IM as Catalyst for the Synthesis of Eugenyl Acetate in Solvent-Free Acetylation. *Appl. Biochem. Biotechnol.* **2015**, *176*, 782–795. [CrossRef] [PubMed]
6. Pandey, S.K.; Tandon, S.; Ahmad, A.; Singh, A.K.; Tripathi, A.K. Structure—Activity relationships of monoterpenes and acetyl derivatives against Aedes aegypti (Diptera: Culicidae) larvae. *Pest Manag. Sci.* **2013**, *69*, 1235–1238. [CrossRef] [PubMed]
7. Charan Raja, M.R.; Velappan, A.B.; Chellappan, D.; Debnath, J.; Mahapatra, S.K. Eugenol derived immunomodulatory molecules against visceral leishmaniasis. *Eur. J. Med. Chem.* **2017**, *139*, 503–518. [CrossRef]

8. Giovannini, P.P.; Sacchetti, G.; Catani, M.; Massi, A.; Tacchini, M.; De Oliveira, D.; Lerin, L.A. Continuous production of eugenol esters using enzymatic bed microreactors and an evaluation of the products as antifungal agents. *Flavour Fragr. J.* **2019**, *34*, 201–210. [CrossRef]
9. Da Siva, F.F.M.; Monte, F.J.Q.; de Lemos, T.L.G.; do Nascimento, P.G.G.; de Medeiros Costa, A.K.; Paiva, L.M.M. Eugenol derivatives: Synthesis, characterization, and evaluation of antibacterial and antioxidant activities. *Chem. Cent. J.* **2018**, *12*, 34. [CrossRef]
10. Yadav, G.D.; Yadav, A.R. Insight into esterification of eugenol to eugenol benzoate using a solid super acidic modified zirconia catalyst UDCaT-5. *Chem. Eng. J.* **2012**, *192*, 146–155. [CrossRef]
11. Manan, F.M.A.; Attan, N.; Zakaria, Z.; Keyon, A.S.A.; Wahab, R.A. Enzymatic esterification of eugenol and benzoic acid by a novel chitosan-chitin nanowhiskers supported Rhizomucor miehei lipase: Process optimization and kinetic assessments. *Enzyme Microb. Technol.* **2018**, *108*, 42–52. [CrossRef] [PubMed]
12. Manan, F.M.A.; Rahman, I.N.A.; Marzuki, N.H.C.; Mahat, N.A.; Huyop, F.; Wahab, R.A. Statistical modelling of eugenol benzoate synthesis using Rhizomucor miehei lipase reinforced nanobioconjugates. *Process Biochem.* **2016**, *51*, 249–262. [CrossRef]
13. Cansian, R.L.; Vanin, A.B.; Orlando, T.; Piazza, S.P.; Puton, B.M.S.; Cardoso, R.I.; Gonçalves, I.L.; Honaiser, T.C.; Paroul, N.; Oliveira, D. Toxicity of clove essential oil and its ester eugenyl acetate against Artemia salina. *Braz. J. Biol.* **2017**, *77*, 155–161. [CrossRef] [PubMed]
14. Teixeira, R.R.; Gazolla, P.A.R.; da Silva, A.M.; Borsodi, M.P.G.; Bergmann, B.R.; Ferreira, R.S.; Vaz, B.G.; Vasconcelos, G.A.; Lima, W.P. Synthesis and leishmanicidal activity of eugenol derivatives bearing. *Eur. J. Med. Chem.* **2018**, *146*, 274–286. [CrossRef] [PubMed]
15. Slamenová, D.; Horváthová, E.; Wsólová, L.; Sramková, M.; Navarová, J. Investigation of anti-oxidative, cytotoxic, DNA-damaging and DNA-protective effects of plant volatiles eugenol and borneol in human-derived HepG2, Caco-2 and VH10 cell lines. *Mutat. Res.* **2009**, *677*, 46–52. [CrossRef]
16. Sadeghian, H.; Seyedi, S.M.; Saberi, M.R.; Arghiani, Z.; Riazi, M. Design and synthesis of eugenol derivatives, as potent 15-lipoxygenase inhibitors. *Bioorg. Med. Chem.* **2008**, *16*, 890–901. [CrossRef]
17. Tischer, J.S.; Possan, H.; Luiz, J.; Malagutti, N.B.; Matello, R.; Valério, A.; Dalmagro, J.; De Oliveira, D.; Oliveira, J.V. Synthesis of eugenyl acetate through heterogeneous catalysis. *J. Essent. Oil Res.* **2019**, *31*, 312–318. [CrossRef]
18. Chiaradia, V.; Paroul, N.; Cansian, R.L.; Júnior, C.V.; Detofol, M.R.; Lerin, L.A.; Oliveira, J.V.; Oliveira, D. Synthesis of eugenol esters by lipase-catalyzed reaction in solvent-free system. *Appl. Biochem. Biotechnol.* **2012**, *168*, 742–751. [CrossRef]
19. Narkhede, N.; Patel, A.; Singh, S. Mono lacunary phosphomolybdate supported on MCM-41: Synthesis, characterization and solvent free aerobic oxidation of alkenes and alcohols. *Dalton Trans.* **2014**, *43*, 2512–2520. [CrossRef]
20. Nascimento, L.A.S.; Angélica, R.S.; Costa, C.E.F.; Zamian, J.R.; Rocha Filho, G.N. Conversion of waste produced by the deodorization of palm oil as feedstock for the production of biodiesel using a catalyst prepared from waste material. *Bioresour. Technol.* **2011**, *102*, 8314–8317. [CrossRef]
21. Nascimento, L.A.S.; Angélica, R.S.; Costa, C.E.F.; Zamian, J.R.; Rocha Filho, G.N. Comparative study between catalysts for esterification prepared from kaolins. *Appl. Clay Sci.* **2011**, *51*, 267–273. [CrossRef]
22. Pires, L.H.O.; Oliveira, A.N.; Monteiro Junior, O.V.; Angélica, R.S.; Costa, C.E.F.; Zamian, J.R.; Nascimento, L.A.S.; Rocha Filho, G.N. Esterification of a waste produced from the palm oil industry over 12-tungstophosforic acid supported on kaolin waste and mesoporous materials. *Appl. Catal. B Environ.* **2014**, *160–161*, 122–128. [CrossRef]
23. Oliveira, A.N.; da Costa, L.R.S.; Pires, L.H.O.; Nascimento, L.A.S.; Angélica, R.S.; Da Costa, C.E.F.; Zamian, J.R.; Da Rocha Filho, G.N. Microwave-assisted preparation of a new esterification catalyst from wasted *flint* kaolin. *Fuel* **2013**, *103*, 626–631. [CrossRef]
24. Oliveira, A.N.; Lima, E.T.L.; Oliveira, D.T.; Andrade, E.H.A.; Angélica, R.S.; Costa, C.E.F.; Rocha Filho, G.N.; Costa, F.F.; Luque, R.; Nascimento, L.A.S. Acetylation of Eugenol over 12-Molybdophosphoric Acid Anchored in Mesoporous Silicate Support Synthesized from *Flint* Kaolin. *Materials* **2019**, *12*, 2995. [CrossRef] [PubMed]
25. Nascimento, L.A.S.; Tito, L.M.Z.; Angélica, R.S.; Costa, C.E.F.; Zamian, J.R.; Rocha Filho, G.N. Esterification of oleic acid over solid acid catalysts prepared from Amazon *flint* kaolin. *Appl. Catal. B Environ.* **2011**, *101*, 495–503. [CrossRef]

26. Carmo, A.C.; de Souza, L.K.C.; da Costa, C.E.F.; Longo, E.; Zamian, J.R.; da Rocha Filho, G.N. Production of biodiesel by esterification of palmitic acid over mesoporous aluminosilicate Al-MCM-41. *Fuel* **2009**, *88*, 461–468. [CrossRef]
27. Oliveira, A.N.; Lima, M.A.B.; Pires, L.H.O.; Silva, M.R.; Luz, P.T.S.; Angélica, R.S.; Rocha Filho, G.N.; Costa, C.E.F.; Luque, R.; Nascimento, L.A.S. Bentonites Modified with Phosphomolybdic Heteropolyacid (HPMo) for Biowaste to Biofuel Production. *Materials* **2019**, *12*, 1431. [CrossRef]
28. Lima, E.T.L.; Queiroz, L.S.; de Pires, L.H.O.; Angélica, R.S.; Costa, C.E.F.; Zamian, J.R.; Rocha Filho, G.N.; Luque, R.; Nascimento, L.A.S. Valorization of Mining Waste in the Synthesis of Organofunctionalized Aluminosilicates for the Esterification of Waste from Palm Oil Deodorization. *ACS Sustain. Chem. Eng.* **2019**, *7*, 7543–7551. [CrossRef]
29. Queiroz, R.M.; Pires, L.H.O.; de Souza, R.C.P.; Zamian, J.R.; de Souza, A.G.; da Rocha Filho, G.N.; da Costa, C.E.F. Thermal characterization of hydrotalcite used in the transesterification of soybean oil. *J. Therm. Anal. Calorim.* **2009**, *97*, 163–166. [CrossRef]
30. Coral, N.; Rodrigues, E.; Rumjanek, V.; Emmerson, C. Soybean biodiesel methyl esters, free glycerin and acid number quantification by 1 H nuclear magnetic resonance spectroscopy. *Magn. Reson. Chem.* **2013**, *51*, 69–71. [CrossRef]
31. Laroque, D.A.; Loss, R.A.; Silva, M.J.A.; Pereira, G.N.; Valerio, A.; Hense, H.; de Oliveira, D.; Oliveira, V. Synthesis of Eugenyl Acetate in Solvent-Free Acetylation: Process Optimization and Kinetic Evaluation. *J. Chem. Eng. Process Technol.* **2015**, *6*, 4–11. [CrossRef]
32. Lerin, L.A.; Catani, M.; Oliveira, D.; Massi, A.; Bortolini, O.; Cavazzini, A.; Giovannini, P.P. Continuous ion-exchange resin catalysed esterification of eugenol for the optimized production of eugenyl acetate using a packed bed microreactor. *RSC Adv.* **2015**, *5*, 76898–76903. [CrossRef]
33. De Souza, L.K.C.; Pardauil, J.J.R.; Zamian, J.R.; Geraldo, N.; Filho, R.; Barrado, C.M.; Angélica, R.S.; Carlos, E.F. Rapid synthesis and characterization of CeMCM-41. *Powder Technol.* **2012**, *229*, 1–6. [CrossRef]
34. Pires, L.H.O.; Queiroz, R.M.; Souza, R.P.; Carlos, E.F.; Zamian, J.R.; Weber, I.T.; Geraldo, N.; Filho, R. Synthesis and characterization of spherical Tb-MCM-41. *J. Alloys Compd.* **2010**, *490*, 667–671. [CrossRef]
35. De Souza, L.K.C.; Pardauil, J.J.R.; Zamian, J.R.; da Rocha Filho, G.N.; Costa, C.E.F. Influence of the incorporated metal on template removal from MCM-41 type mesoporous materials. *J. Therm. Anal. Calorim.* **2011**, *106*, 355–361. [CrossRef]
36. Zhou, C.; Sun, T.; Gao, Q.; Alshameri, A.; Zhu, P.; Wang, H.; Qiu, X.; Ma, Y.; Yan, C. Synthesis and characterization of ordered mesoporous aluminosilicate molecular sieve from natural halloysite. *J. Taiwan Inst. Chem. Eng.* **2014**, *45*, 1073–1079. [CrossRef]
37. Xie, Y.; Zhang, Y.; Ouyang, J.; Yang, H. Mesoporous material Al-MCM-41 from natural halloysite. *Phys. Chem. Miner.* **2014**, *41*, 497–503. [CrossRef]
38. Santos, E.C.; Costa, L.S.; Oliveira, E.S.; Bessa, R.A.; Freitas, A.D.L.; Oliveira, C.P.; Nascimento, R.F.; Loiola, A.R. Al-MCM-41 synthesized from kaolin via hydrothermal route: Structural characterization and use as an efficient adsorbent of methylene blue. *J. Braz. Chem. Soc.* **2018**, *29*, 2378–2386. [CrossRef]
39. Wang, G.; Wang, Y.; Liu, Y.; Liu, Z.; Guo, Y.; Liu, G.; Yang, Z.; Xu, M.; Wang, L. Synthesis of highly regular mesoporous Al-MCM-41 from metakaolin. *Appl. Clay Sci.* **2009**, *44*, 185–188. [CrossRef]
40. Madhusoodana, C.D.; Kameshima, Y.; Nakajima, A.; Okada, K.; Kogure, T.; MacKenzie, K.J.D. Synthesis of high surface area Al-containing mesoporous silica from calcined and acid leached kaolinites as the precursors. *J. Colloid Interface Sci.* **2006**, *297*, 724–731. [CrossRef]
41. Kang, F.; Wang, Q.; Xiang, S. Synthesis of mesoporous Al-MCM-41 materials using metakaolin as aluminum source. *Mater. Lett.* **2005**, *59*, 1426–1429. [CrossRef]
42. Du, C.; Yang, H. Investigation of the physicochemical aspects from natural kaolin to Al-MCM-41 mesoporous materials. *J. Colloid Interface Sci.* **2012**, *369*, 216–222. [CrossRef] [PubMed]
43. Rocha Junior, C.A.F.; Angélica, R.S.; Neves, R.F. Sinthesis of faujasite-type zeolite: Comparison between processed and *flint* kaolin. *Cerâmica* **2015**, *61*, 259–268. [CrossRef]
44. Carneiro, B.S.; Angélica, R.S.; Scheller, T.; de Castro, E.A.S.; de Neves, R.F. Mineralogical and geochemical characterizations of the hard kaolin from the Capim region, Pará, northern Brazil. *Cerâmica* **2003**, *49*, 237–244. [CrossRef]

45. Sun, C.; Zhang, F.; Wang, X.; Cheng, F. Facile Preparation of Ammonium Molybdophosphate/Al-MCM-41 Composite Material from Natural Clay and Its Use in Cesium Ion Adsorption. *Eur. J. Inorg. Chem.* **2015**, *2015*, 2125–2131. [CrossRef]
46. Sun, C.; Zhang, F.; Li, S.; Cheng, F. Synthesis of SBA-15 encapsulated ammonium molybdophosphate using Qaidam natural clay and its use in cesium ion adsorption. *RSC Adv.* **2015**, *5*, 35453–35460. [CrossRef]
47. Díaz, U.; Brunel, D.; Corma, A. Catalysis using multifunctional organosiliceous hybrid materials. *Chem. Soc. Rev.* **2013**, *42*, 4083–4097. [CrossRef]
48. Ng, E.; Norbayu, S.; Subari, M.; Marie, O.; Mukti, R.R.; Juan, J. Sulfonic acid functionalized MCM-41 as solid acid catalyst for tert -butylation of hydroquinone enhanced by microwave heating. *Appl. Catal. A Gen.* **2013**, *450*, 34–41. [CrossRef]
49. Wang, Y.; Fang, Z.; Zhang, F. Esterification of oleic acid to biodiesel catalyzed by a highly acidic carbonaceous catalyst. *Catal. Today* **2019**, *319*, 172–181. [CrossRef]
50. Boveri, M.; Aguilar-Pliego, J.; Pérez-Pariente, J.; Sastre, E. Optimization of the preparation method of HSO_3-functionalized MCM-41 solid catalysts. *Catal. Today* **2005**, *107–108*, 868–873. [CrossRef]
51. Guo, K.; Han, F.; Arslan, Z.; McComb, J.; Mao, X.; Zhang, R.; Sudarson, S.; Yu, H. Adsorption of Cs from Water on Surface-Modified MCM-41 Mesosilicate. *Water Air Soil Pollut.* **2015**, *226*, 2–9. [CrossRef]
52. Fontes, M.S.B.; Melo, D.M.A.; Costa, C.C.; Braga, R.M.; Melo, M.A.F.; Alves, J.A.B.L.R.; Silva, M.L.P. Effect of different silica sources on textural parameters of molecular sieve MCM-41. *Cerâmica* **2016**, *62*, 85–90. [CrossRef]
53. Kumar, P.; Mal, N.; Oumi, Y.; Yamana, K.; Sano, T. Mesoporous materials prepared using coal fly ash as the silicon and aluminium source. *J. Mater. Chem.* **2001**, *11*, 3285–3290. [CrossRef]
54. Yang, H.; Deng, Y.; Du, C.; Jin, S. Novel synthesis of ordered mesoporous materials Al-MCM-41 from bentonite. *Appl. Clay Sci.* **2010**, *47*, 351–355. [CrossRef]
55. Sing, K. The use of nitrogen adsorption for the characterisation of porous materials. *Colloids Surf. A Physicochem. Eng. Asp.* **2001**, *187–188*, 3–9. [CrossRef]
56. Sing, K.S.W.; Everett, D.H.; Haul, R.A.W.; Moscou, L.; Pierotti, A.R.; Rouquérol, J.; Siemieniewska, T. Reporting Physisorption Data for Gas/Solid Systems including catalysis reporting physisorption data for gas/solid systems with Special Reference to the Determination of Surface Area and Porosity. *Pure Appl. Chem.* **1985**, *57*, 603–619. [CrossRef]
57. Patel, A.; Brahmkhatri, V. Kinetic study of oleic acid esterification over 12-tungstophosphoric acid catalyst anchored to different mesoporous silica supports. *Fuel Process. Technol.* **2013**, *113*, 141–149. [CrossRef]
58. Brahmkhatri, V.; Patel, A. 12-Tungstophosphoric acid anchored to SBA-15: An efficient, environmentally benign reusable catalysts for biodiesel production by esterification of free fatty acids. *Appl. Catal. A Gen.* **2011**, *403*, 161–172. [CrossRef]
59. Ahmed, A.I.; Samra, S.E.; El-Hakam, S.A.; Khder, A.S.; El-Shenawy, H.Z.; El-Yazeed, W.S.A. Characterization of 12-molybdophosphoric acid supported on mesoporous silica MCM-41 and its catalytic performance in the synthesis of hydroquinone diacetate. *Appl. Surf. Sci.* **2013**, *282*, 217–225. [CrossRef]
60. Méndez, F.J.; Llanos, A.; Echeverría, M.; Jáuregui, R.; Villasana, Y.; Díaz, Y.; Liendo-Polanco, G.; Ramos-García, M.A.; Zoltan, T.; Brito, J.L. Mesoporous catalysts based on Keggin-type heteropolyacids supported on MCM-41 and their application in thiophene hydrodesulfurization. *Fuel* **2013**, *110*, 249–258. [CrossRef]
61. Costa, B.O.D.; Legnoverde, M.S.; Lago, C.; Decolatti, H.P.; Querini, C.A. Microporous and Mesoporous Materials Sulfonic functionalized SBA-15 catalysts in the gas phase glycerol dehydration. Thermal stability and catalyst deactivation. *Microporous Mesoporous Mater.* **2016**, *230*, 66–75. [CrossRef]
62. Adam, F.; Kueh, C.W. Phenyl-amino sulfonic solid acid-MCM-41 complex: A highly active and selective catalyst for the synthesis of mono-alkylated products in the solvent free tert-butylation of phenol. *J. Taiwan Inst. Chem. Eng.* **2014**, *45*, 713–723. [CrossRef]
63. Zhang, P.; Wu, H.; Fan, M.; Sun, W.; Jiang, P.; Dong, Y. Direct and postsynthesis of tin-incorporated SBA-15 functionalized with sulfonic acid for efficient biodiesel production. *Fuel* **2019**, *235*, 426–432. [CrossRef]
64. Lima, E.T.L. Síntese de Al-MCM-41 a partir do rejeito do caulim e impregnação com grupo sulfônico para fins catalíticos. 2016, 63f. In *Dissertação (Mestrado em Química)–Programa de Pós-Graduação em Química*; University Federal do Pará: Belém, Brazil, 2016.

65. Campelo, J.M.; Lafont, F.; Marinas, J.M.; Ojeda, M. Studies of catalyst deactivation in methanol conversion with high, medium and small pore silicoaluminophosphates. *Appl. Catal. A Gen.* **2000**, *192*, 85–96. [CrossRef]
66. Gang, L.; Xinzong, L.; Eli, W. Solvent-free esterification catalyzed by surfactant-combined catalysts at room temperature. *New J. Chem.* **2007**, *31*, 348. [CrossRef]
67. Santos, P.; Zabot, G.L.; Meireles, M.A.A.; Mazutti, M.A.; Martínez, J. Synthesis of eugenyl acetate by enzymatic reactions in supercritical carbon dioxide. *Biochem. Eng. J.* **2016**, *114*, 1–9. [CrossRef]
68. Narkhede, N.; Singh, S.; Patel, A. Recent progress on supported polyoxometalates for biodiesel synthesis via esterification and transesterification. *Green Chem.* **2015**, *17*, 89–107. [CrossRef]
69. Wang, A.; Wang, J.; Lu, C.; Xu, M.; Lv, J.; Wu, X. Esterification for biofuel synthesis over an eco-friendly and efficient kaolinite-supported $SO_{4}^{2-}/ZnAl_2O_4$ macroporous solid acid catalyst. *Fuel* **2018**, *234*, 430–440. [CrossRef]
70. Hoo, P.; Abdullah, A.Z. Kinetics Modeling and Mechanism Study for Selective Esterification of Glycerol with Lauric Acid Using 12-Tungstophosphoric Acid Post-Impregnated SBA-15. *Ind. Eng. Chem. Res.* **2015**, *54*, 7852–7858. [CrossRef]
71. Baskaran, Y.; Periyasamy, V.; Carani, A. Investigation of antioxidant, anti-inflammatory and DNA-protective properties of eugenol in thioacetamide-induced liver injury in rats. *Toxicology* **2010**, *268*, 204–212. [CrossRef]
72. Devi, K.P.; Nisha, S.A.; Sakthivel, R.; Pandian, S.K. Eugenol (an essential oil of clove) acts as an antibacterial agent against Salmonella typhi by disrupting the cellular membrane. *J. Ethnopharmacol.* **2010**, *130*, 107–115. [CrossRef] [PubMed]
73. Santin, J.R.; Lemos, M.; Klein-Júnior, L.C.; Machado, I.D.; Costa, P.; De Oliveira, A.P.; Tilia, C.; De Souza, J.P.; De Sousa, J.P.B.; Bastos, J.K.; et al. Gastroprotective activity of essential oil of the Syzygium aromaticum and its major component eugenol in different animal models. *Naunyn-Schmiedeberg's Arch. Pharmacol.* **2011**, *383*, 149–158. [CrossRef]
74. Rodrigues, T.; Fernandes, A., Jr.; Sousa, J.; Bastos, J.; Sforcin, J. In vitro and in vivo effects of clove on pro-inflammatory cytokines production by macrophages. *Nat. Prod. Res.* **2009**, *23*, 319–326. [CrossRef] [PubMed]
75. Chaibakhsh, N.; Basri, M.; Anuar, S.H.M.; Rahman, M.B.A.; Rezayee, M. Optimization of enzymatic synthesis of eugenol ester using statistical approaches. *Biocatal. Agric. Biotechnol.* **2012**, *1*, 226–231. [CrossRef]
76. Affonso, R.S.; Lessa, B.; Slana, G.B.C.A.; Barboza, L.L.; de Almeida, F.V.; de Souza, F.R.; França, T.C.C. Quantification and Characterization of the Main Components of the Ethanolic Extract of Indian Cloves, Syzygium aromaticum [l] Mer. et Perry. *Rev. Virtual Quim.* **2014**, *6*, 1316–1331. [CrossRef]
77. Brunauer, S.; Emmett, P.H.; Teller, E. Adsorption of Gases in Multimolecular Layers. *J. Am. Chem. Soc.* **1938**, *60*, 309–319. [CrossRef]
78. Reddy, C.R.; Nagendrappa, G.; Prakash, B.S.J. Surface acidity study of Mn^+-montmorillonite clay catalysts by FT-IR spectroscopy: Correlation with esterification activity. *Catal. Commun.* **2007**, *8*, 241–246. [CrossRef]
79. Reddy, C.R.; Bhat, Y.S.; Nagendrappa, G.; Prakash, B.S.J. Brønsted and Lewis acidity of modified montmorillonite clay catalysts determined by FT-IR spectroscopy. *Catal. Today* **2009**, *141*, 157–160. [CrossRef]
80. NIST NIST (2011) National Institute of Standard and Technology (2011) NIST Standard Reference Database Number 69. Available online: http://webbook.nist.gov/ (accessed on 25 November 2018).

© 2020 by the authors. Licensee MDPI, Basel, Switzerland. This article is an open access article distributed under the terms and conditions of the Creative Commons Attribution (CC BY) license (http://creativecommons.org/licenses/by/4.0/).

Article

One-Pot Alcoholysis of the Lignocellulosic *Eucalyptus nitens* Biomass to *n*-Butyl Levulinate, a Valuable Additive for Diesel Motor Fuel

Claudia Antonetti [1], Samuele Gori [1], Domenico Licursi [1,*], Gianluca Pasini [2], Stefano Frigo [2], Mar López [3], Juan Carlos Parajó [3] and Anna Maria Raspolli Galletti [1,*]

1. Department of Chemistry and Industrial Chemistry, University of Pisa, Via Giuseppe Moruzzi 13, 56124 Pisa, Italy; claudia.antonetti@unipi.it (C.A.); samuele.gori94@gmail.com (S.G.)
2. Department of Energy, Systems, Territory and Construction Engineering, University of Pisa, Largo Lucio Lazzarino, 56122 Pisa, Italy; gianluca.pasini@for.unipi.it (G.P.); stefano.frigo@unipi.it (S.F.)
3. Department of Chemical Engineering, Faculty of Science, University of Vigo (Campus Ourense), As Lagoas, 32004 Ourense, Spain; marlopezr@uvigo.es (M.L.); jcparajo@uvigo.es (J.C.P.)
* Correspondence: domenico.licursi@unipi.it (D.L.); anna.maria.raspolli.galletti@unipi.it (A.M.R.G.); Tel.:+39-050-2210543 (D.L.); +39-050-2219290 (A.M.R.G.)

Received: 17 April 2020; Accepted: 3 May 2020; Published: 6 May 2020

Abstract: The present investigation represents a concrete example of complete valorization of *Eucalyptus nitens* biomass, in the framework of the circular economy. Autohydrolyzed-delignified *Eucalyptus nitens* was employed as a cheap cellulose-rich feedstock in the direct alcoholysis to *n*-butyl levulinate, adopting *n*-butanol as green reagent/reaction medium, very dilute sulfuric acid as a homogeneous catalyst, and different heating systems. The effect of the main reaction parameters to give *n*-butyl levulinate was investigated to check the feasibility of this reaction and identify the coarse ranges of the main operating variables of greater relevance. High *n*-butyl levulinate molar yields (35–40 mol%) were achieved under microwave and traditional heating, even using a very high biomass loading (20 wt%), an eligible aspect from the perspective of the high gravity approach. The possibility of reprocessing the reaction mixture deriving from the optimized experiment by the addition of fresh biomass was evaluated, achieving the maximum *n*-butyl levulinate concentration of about 85 g/L after only one microwave reprocessing of the mother liquor, the highest value hitherto reported starting from real biomass. The alcoholysis reaction was further optimized by Response Surface Methodology, setting a Face-Centered Central Composite Design, which was experimentally validated at the optimal operating conditions for the *n*-butyl levulinate production. Finally, a preliminary study of diesel engine performances and emissions for a model mixture with analogous composition to that produced from the butanolysis reaction was performed, confirming its potential application as an additive for diesel fuel, without separation of each component.

Keywords: *n*-butyl levulinate; alcoholysis; butanolysis; *Eucalyptus nitens*; microwaves; biorefinery; diesel blends

1. Introduction

Levulinic acid (LA) is a biomass-derived platform chemical which has attracted increasing interest in recent years due to the possibility to be converted into added-value derivatives, such as biofuels, fragrances, solvents, pharmaceuticals, and plasticizers [1], thus justifying the increasing worldwide market demand for LA production [2]. LA is traditionally produced in water medium via dehydration of C6 sugars through the formation of 5-hydroxymethylfurfural as the main reaction intermediate, the overall reaction occurring in the presence of a suitable acid catalyst [3–7]. Among the LA-derived platforms chemicals, alkyl levulinates appear significantly attractive due to their potential applications developed in recent years for the global market scenario, such as fuel blending additives for diesel/gasoline, and as intermediates for the synthesis of valuable polymers, perfumes, and flavoring formulations [8,9]. Levulinates can be synthesized by the esterification of pure LA with a simple equilibrium reaction, requiring a mild acid catalysis/reaction conditions, and generally affording very high yields towards the desired ester products. Both the reduced number of process units and the enhanced performances of new technological solutions, such as the reactive distillation, should allow significant improvements in the economics of the esterification process [10–12]. However, despite these ascertained potentials, the catalysis issue can be further improved, taking into account both the synthetic strategy and the adopted feedstock. Up to now, much work has been done on the synthesis of methyl and ethyl levulinates, which were recognized as effective additives for diesel and biodiesel transportation fuels, showing excellent performances, including non-toxicity, high lubricity and good flashpoint stability and flow properties under cold conditions [13–15]. In addition, the conversion of these short-chain alkyl levulinates into more added-value bio-products, such as γ-valerolactone, is preferred respect to that of LA due to the improved selectivity. Moreover, the hydrogenation of alkyl levulinates with short alkyl chains facilitates down-stream processing as separation of alcohol in the final step is easier and cheaper, compared to water [16]. The synthesis of alkyl levulinates was carried out in the presence of homogeneous or, more advantageously, heterogeneous catalysts due to their easy recovery from the reaction mixture, starting from pure LA or expensive pure model precursors, such as C6 carbohydrates (glucose, fructose, and clean cellulose), C5 derivatives (furfuryl alcohol), and even real lignocellulosic biomasses [17,18]. The one-pot synthesis of these levulinates directly from monosaccharides, polysaccharides, and, above all, starting from lignocellulosic biomass, has gained more interest due to the low cost of these feedstocks, and the feasibility of this approach was demonstrated, in particular for the biomass alcoholysis to ethyl levulinate [19]. A key advantage of the direct alcoholysis is represented by the limited formation of undesired furanic products, named *humins*, when using alcohol (instead of water) as the solvent for biomass conversion [20,21]. On the other hand, the yields of levulinate esters from real biomass are generally lower than those obtained from pure model compounds due to the usually higher recalcitrance of the former [18], and to the increased formation of reaction by-products, such as formates, HMF ethers and, above all, dialkyl ethers, originating from the alcohol dehydration [20,21]. Differently, *n*-butyl levulinate (BL) was less studied, but its use as an efficient fuel additive was already demonstrated [22], resulting in a more promising diesel additive than EL [18,23]. In addition, *n*-butanol (*n*-BuOH) is a green reagent/solvent, being obtainable by fermentation and also by catalytic conversion of bio-ethanol [24], thus further justifying the interest towards the sustainable production of BL. Regarding the possible pathways for BL production, as previously stated for methyl and ethyl levulinates, it can also be obtained with a two-steps process from C5 or C6 carbohydrates or their conversion products (Pathway A or Pathway B_1, respectively, in Figure 1) or, more advantageously, with a one-pot approach from C6 carbohydrates (Pathway B_2, Figure 1). In the first case, furfuryl alcohol or LA (from hemicellulose and cellulose fractions, respectively) must be synthesized in the first step, recovered, and properly purified before the subsequent stage, consisting of acid alcoholysis or esterification, respectively. The C5 route (Pathway A, Figure 1) is a three-step process consisting of: (1) acid-catalyzed hydrolysis of the hemicellulose fraction to simpler C5 sugars and their dehydration to furfural; (2) hydrogenation of furfural over a suitable catalyst to furfuryl alcohol; (3) acid alcoholysis of furfuryl alcohol to BL, occurring in the presence of strong acid catalysts [25]. The C5 route was investigated in the literature adopting furfuryl alcohol as starting feedstock, in the presence of heterogeneous catalysts, due to their easier separation from the liquid reaction mixture [26–29]. On the other hand, BL synthesis through the C6 route (Pathway B_1, Figure 1) provides the hydrolysis of the C6 carbohydrates to LA, followed by its esterification in *n*-BuOH, and both steps occur in the presence of a suitable acid catalyst.

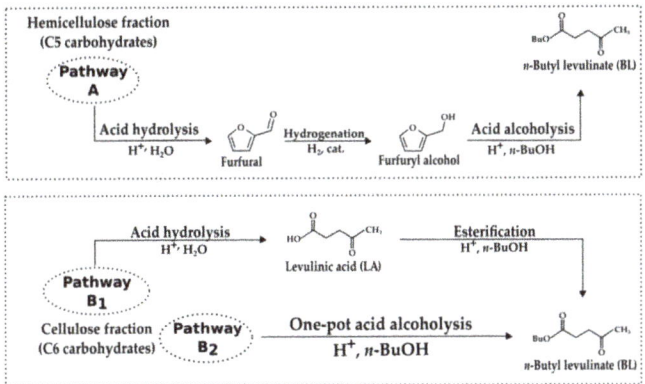

Figure 1. C5 and C6 sugar-based routes to *n*-butyl levulinate.

Regarding this C6 route, in the literature, it is possible to find many BL synthesis from pure LA, the intermediate compound, usually preferring the use of heterogeneous catalysts, achieving excellent yields (>90 mol%) under sustainable reaction conditions [9,30,31]. On the contrary, BL synthesis from C6 carbohydrates was not exploited with the same emphasis, although this approach should result very attractive from the industrial perspective if realized in a single step without any intermediate purification procedures (Pathway B$_2$, Figure 1), thus decreasing the BL production cost. In this context, some authors have reported the one-pot butanolysis of microcrystalline cellulose to BL [32–41], which is already very difficult to achieve due to its recalcitrance to the solubilization/conversion, while the butanolysis of the real biomass, which includes lignin as a further recalcitrant component, is even unexplored. In this regard, a simplified scheme of the C6 fraction butanolysis pathway is shown in Figure 2.

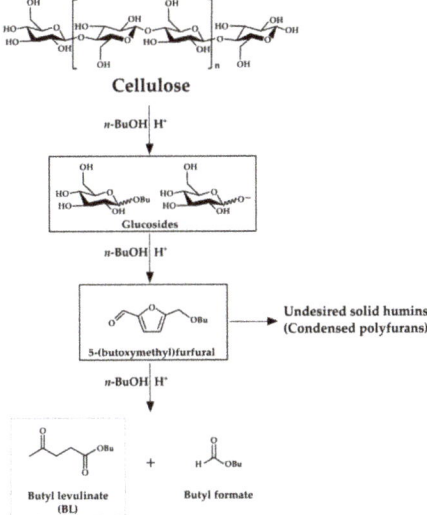

Figure 2. C6 fraction butanolysis pathway starting from cellulose feedstock (adapted from [40]).

Butanolysis of the C6 fraction is a complex pathway, which involves the formation of many reactive species, in particular butyl glucosides and furanic derivatives as the main reaction intermediates, in addition to butyl formate (BF) as the main reaction co-product. Furanic intermediates are very reactive species, which could condense to solid insoluble polyfurans, the humins [40]. The first step of the butanolysis process consists of the depolymerization of cellulose chains to form glucosides, followed by the subsequent formation of furan derivatives, whereas the final step involves the conversion of the furanic intermediates to BL, and all these steps occur in the presence of an acid catalyst [40]. When the above reaction is performed adopting a real solid lignocellulosic biomass, the use of homogeneous catalysts is the best choice. Some typical drawbacks, such as the possible corrosion of the equipment, and the recovery of the acid catalyst, need to be further improved by adopting very low acid concentrations and more technological work-up solutions. Moreover, the use of a very low acid concentration in the alcoholysis reaction, which helps to minimize the corrosion of the equipment, should also control the formation extent of by-products, in particular the dialkyl ether [8,14]. In this context, it is noteworthy the work of Démolis et al. [37], who achieved the highest BL molar yield of about 50 mol%, working in an autoclave at 200 °C for 30 min, adopting pure cellulose as starting feedstock (2.4 wt%), with a very low concentration of H_2SO_4 (0.6 wt%). However, this good BL yield, although academically interesting, was obtained with a low starting cellulose loading, which should represent significant limitations for the development on the intensified industrial scale.

Definitely, at this state-of-the-art, the main bottlenecks of the published works are related to the adoption of (1) model compounds as starting substrates instead of the cheaper and largely available real biomasses, and (2) low substrate loading, which is not a limit for an academic investigation, in a preliminary phase, but it is certainly for the next industrial scale-up. Therefore, the resolution of both these aspects is fundamental for the BL development towards the biofuel market, and this work contributes to filling this gap. In this context, wood is the most abundant type of lignocellulosic biomass and, more in detail, *Eucalyptus* is a widespread, fast-growing, and widely distributed species, which shows a good adaptation to grow in zones with a high probability of freezing and affording decreased susceptibility to diseases [42–44]. It already shows an interesting potential in many industrial fields, as in the paper-making production, where it is already used as a valuable and cheap fiber source. Moreover, it is an ideal energy crop, thanks to its high yield, low energy input for production, low cost, minimal contents of contaminants, and low nutrient requirements. From a different perspective, it may represent a promising feedstock for many biochemical conversion processes, given its high content of C5 and C6 carbohydrates (about 60 wt%) [42–44]. The use of this feedstock is particularly advantageous if the complete fractionation and the successive valorization of each component are achieved, according to the perspective of an integrated biorefinery [45]. In this context, the aim of the present work is the complete exploitation of *Eucalyptus nitens* biomass. For this purpose, pre-treated autohydrolyzed-delignified wood (ADW) *Eucalyptus nitens* was obtained from a first autohydrolysis treatment of the starting raw biomass in order to remove and exploit hemicellulose and water-soluble extractives, followed by a second step of delignification on the resulting solid through the HCl-catalyzed acetic acid treatment (Acetosolv method). The recovered cellulose-rich feedstock was employed for the one-pot production of BL in *n*-BuOH, adopting microwave (MW) and/or traditional (TR) heating, in the presence of very dilute sulfuric acid as a homogeneous catalyst. MW heating represents an important tool because it can reduce reaction time and energy consumption, thus improving the efficiency of the process [46,47]. In the specific case of LA esterification, remarkable thermal (kinetic) advantages of MW towards this reaction were already reported by Ahmad et al. [48]. The choice of H_2SO_4 as the acid catalyst was done taking into account its promising catalytic performances in the alcoholysis reaction to methyl and ethyl levulinates [49,50], while other acid catalysts, such as HCl or H_3PO_4, resulted less active, for example in the case of the one-pot reaction from cellulose to ethyl levulinate [49]. The effects of the main reaction parameters, temperature, reaction time, and acid concentration were investigated by a traditional One-Factor-at-a-Time (OFAT) approach and further optimized by Response Surface Methodology (RSM), developing a Face-Centered Central Composite Design (FCCD), from the perspective of developing the BL process intensification. Finally, a preliminary study of diesel engine performances and emissions for a model mixture with analogous

2. Results and Discussion

2.1. Characterization of the Eucalyptus nitens Samples

The chemical composition of the starting untreated *Eucalyptus nitens* was as follows: 42.0 wt% of cellulose, 14.5 wt% of hemicellulose, 21.4 wt% of Klason lignin, and 22.1 wt% of unidentified compounds (including acid-soluble lignin, ash, extractives, waxes). After carrying out the autohydrolysis and Acetosolv pretreatments, the mass yield of the ADW *Eucalyptus nitens* sample was 45.0 wt% of the starting raw biomass, and its chemical composition resulted to be as follows: 85.0 wt% cellulose, 2.0 wt% hemicellulose, 4.1 wt% Klason lignin, 4.8 wt% acetyl groups, and 4.1 wt% of unidentified other compounds. The compositional analysis of untreated and ADW *Eucalyptus nitens* samples confirms the effective enrichment in cellulose and the depletion in hemicellulose and lignin as a consequence of the chemical pre-treatments [51].

XRD analysis of the untreated and ADW *Eucalyptus nitens* samples was attained in order to estimate the crystallinity index (CI) of the cellulose fraction, a paramount parameter for understanding the behavior of biomass to the subsequent butanolysis reaction, achievable under an appropriate severity degree [52]. The XRD spectra of the starting untreated and ADW *Eucalyptus nitens* biomasses are reported in Figure 3. Here, deconvoluted curves were reported, including that due to amorphous cellulose (at about $2\theta = 21.5°$) and those related to the crystalline planes, with Miller indices of 101, 10$\bar{1}$, 002 e 040.

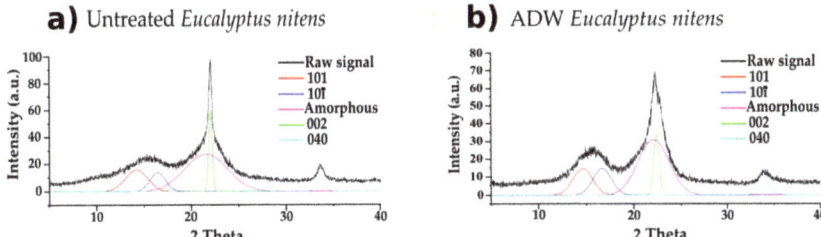

Figure 3. XRD spectra of (**a**) untreated and (**b**) autohydrolyzed-delignified wood (ADW) *Eucalyptus nitens* samples.

A higher crystallinity degree was obtained for the ADW sample rather than for the untreated one (46.8 versus 43.3%, respectively), ascribed to the partial removal of both lignin and hemicellulose fractions for the ADW sample as a consequence of the pre-treatment, leading to greater exposure of the crystalline cellulose fraction [53]. Besides, autohydrolysis pre-treatment allowed the preferential removal of the amorphous component of the cellulose, leaving almost unchanged the crystalline portion [54].

FT–IR characterization of the untreated and ADW *Eucalyptus nitens* samples was also carried out, and the acquired spectra are reported in Figure 4.

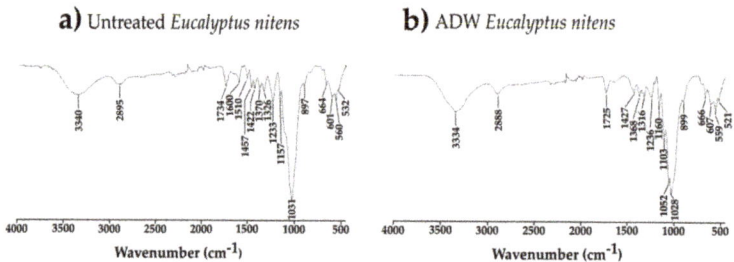

Figure 4. FT–IR spectra of (**a**) untreated and (**b**) ADW *Eucalyptus nitens* samples.

In the IR spectrum of the untreated *Eucalyptus nitens* sample, typical bands of biomass macro-components, cellulose, hemicellulose, and lignin derivatives, are detected, such as that at about 3400 cm^{-1}, assigned to the O-H stretching, and that at about 2900 cm^{-1}, due to the C-H stretching. Moreover, the absorption band at about 1730 cm^{-1} is assigned to the C=O stretching of ester bonds, such as acetyl derivatives, while those at 1600 cm^{-1} and 1510 cm^{-1} indicate the presence of C=C ring vibrations, which are typical of lignin units [55,56]. In the region between 1500 and 1300 cm^{-1}, absorption bands ascribed to the bending of the O-H bonds and the vibrations of the methyl and methylene groups of both lignin and cellulose are present. The absorption bands between 1300 and 1200 cm^{-1} are due to the stretching of the C-O bonds of the alcoholic, phenolic, and carboxyl groups. The shoulder at about 1160 cm^{-1} can be assigned to the stretching of the C-O-C bond of the hemicellulose and cellulose, while the absorption bands at about 1030 cm^{-1} and that at about 900 cm^{-1} are due to the stretching of the C-O-C β-glycosidic bonds of the cellulose [55,56]. Regarding the IR spectrum of ADW *Eucalyptus nitens* biomass, the absorption bands of lignin rings at 1600 cm^{-1} and 1510 cm^{-1} are absent, thus confirming the efficacy of the Organosolv treatment. In addition, a new absorption band is present at about 1050 cm^{-1}, which is uniquely assigned to the C–O stretching of the cellulose [57], thus indirectly confirming the occurred cellulose enrichment for the ADW *Eucalyptus nitens* sample. The other absorption bands are similar to those discussed for the untreated *Eucalyptus nitens* biomass.

2.2. Univariate Optimization: OFAT Approach

After having demonstrated the occurred cellulose enrichment of the ADW *Eucalyptus nitens*, this biomass was our preferred choice for performing the next one-pot butanolysis to BL, thus further developing the biorefinery concept of this biomass. For this purpose, alcoholysis of ADW *Eucalyptus nitens* (range of biomass loading: 7–20 wt%) in *n*-BuOH was preliminarily investigated by a traditional One-Factor-at-a-Time (OFAT) approach, employing both MW and TR heating, in the presence of 1.2 wt% H$_2$SO$_4$ (Table 1). Starting from the published results [8,36], in the beginning, the biomass loading of 7 wt% and the temperature of 190 °C, under MW heating, were selected for studying the behavior of the reaction (Runs 1–3, Table 1). At the increase of the reaction time, the BL molar yield raised to 42 mol% after 15 min. The extension of the reaction time did not affect the BL molar yield, which was stable at 42 mol%. Under the optimized reaction conditions (MW, 190 °C, 15 min), the comparison between ADW and untreated *Eucalyptus nitens* biomass was investigated (Runs 2 and 4, Table 1): the employment of the starting crude *Eucalyptus nitens* without any pre-treatment allowed us to obtain the same BL molar yield of 42 mol% as the corresponding ADW sample, but the BL concentrations in the final reaction mixtures were 23 g/L for the ADW sample against 12 g/L for the crude *Eucalyptus nitens* due to the higher cellulose content in the ADW biomass. Finally, a further test employing TR heating was carried out (Run 5, Table 1) employing the ADW *Eucalyptus nitens* wood as substrate: after 120 min, the BL molar yield of 49 mol% was obtained together with the BL concentration of 27 g/L, demonstrating that analogous promising results can be also achieved with TR heating, even if a longer reaction time was necessary.

Table 1. One-pot butanolysis of the untreated or ADW *Eucalyptus nitens* to *n*-butyl levulinate (BL), adopting microwave (MW), or traditional (TR) heating. Reaction conditions: biomass loading 7 wt%, 190 °C, H_2SO_4 1.2 wt%.

Run	Biomass	Heating	Time (min)	BL Yield (mol%)	BL Conc. (g/L)
1	ADW	MW	10	32	18
2	ADW	MW	15	42	23
3	ADW	MW	30	42	23
4	Untreated	MW	15	42	12
5	ADW	TR	120	49	27

Taking into account the low cost of the starting biomass, it is more important to achieve high BL concentrations in the final mixture rather than to maximize the BL molar yield respect to cellulose fraction present in the starting biomass, making the entire process economically convenient due to the significant reduction of the purification cost, from the perspective of the high gravity approach [58]. On this basis, the biomass loading was increased to 14 and 20 wt% and the obtained results adopting MW heating, working at 190 °C, for 15 min, in the presence of H_2SO_4 1.2 wt%, are shown in Figure 5.

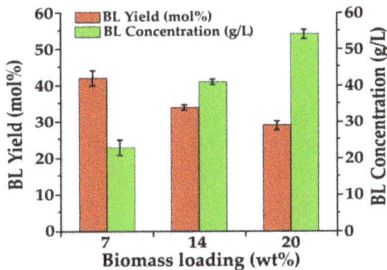

Figure 5. One-pot MW-assisted butanolysis of the ADW *Eucalyptus nitens* sample to BL, adopting different biomass loadings (7, 14, and 20 wt%). Reaction conditions: 190 °C, 15 min, H_2SO_4 1.2 wt%, MW heating.

The increase of the initial biomass loading caused the decrease of BL molar yield as expected considering that, when a higher initial biomass loading is employed, adopting the same amount of catalyst, not only the catalyst/biomass weight ratio decreases, but also the mixture mixing can become more difficult, working in slurry phase. However, at the increase of initial biomass loading, the decrease of BL molar yield was not significant and it was associated with a huge increase of BL concentrations, highlighting the effectiveness of the high gravity approach. The same reactions were also carried out under TR heating to confirm the feasibility of this reaction on a larger scale, and the comparison between the two systems is shown in Table 2.

Table 2. One-pot butanolysis of the ADW *Eucalyptus nitens* to BL, with different biomass loadings (7, 14, and 20 wt%), adopting MW and TR heating systems. Reaction conditions: 190 °C, H_2SO_4 1.2 wt%.

Run	Biomass Loading (wt%)	Heating	Time (min)	BL Yield (mol%)	BL Conc. (g/L)
6	ADW 7 wt%	TR	120	49	27
7	ADW 7 wt%	MW	15	42	23
8	ADW 14 wt%	TR	120	44	53
9	ADW 14 wt%	MW	15	34	41
10	ADW 20 wt%	TR	120	37	69
11	ADW 20 wt%	MW	15	29	54

The shift from the MW to the TR heating system was demonstrated, in the latter case requiring much longer reaction times to get comparable BL molar yields. As already achieved for the MW heating, also for TR, the systematic decrease of the BL molar yield, occurring with the increase of the initial biomass loading, may be due to the increase of the substrate/catalyst ratio, leading to an insufficient amount of catalyst, and to the inefficient agitation of the reaction slurry.

At this level of investigation, the above tests with TR heating confirm the MW data, further justifying and claiming our high gravity approach. Good results were achieved with the biomass loading of 20 wt% (Runs 10 and 11, Table 2), and therefore, 25 wt% of biomass loading was also tested, always under TR heating. Unfortunately, in this last case, although the decrease of BL molar yield (35 mol%) was not significant with a related very high BL concentration (87 g/L), considerable practical difficulties were encountered in filtering and recovering the liquid phase. For this reason, even in the case of TR heating, the best result, in the application perspective, is obtained with the biomass loading of 20 wt%. The achieved results are very interesting because, up to now, BL molar yield higher than 30 mol%, corresponding to the best BL concentration of about 70 g/L, with the initial biomass loading of 20 wt%, has never been ascertained under TR heating, opening the way towards the industrial adoption of this approach for BL synthesis.

On the basis of the promising results obtained with the biomass loading of 20 wt%, the best alcoholysis test under TR heating (Run 10, Table 2) was further investigated, adopting lower reaction times (30 and 60 min), to gather more information about the kinetics. These new data are reported in Figure 6. This figure shows that the most significant improvement, in terms of BL molar yield, was achieved already after 60 min. Lower reaction times (30 min) are not sufficient for the complete conversion of the reaction intermediates to BL, while higher ones (120 min) are not advantageous, not leading to further raise of BL molar yield. In terms of BL concentration, the increase is remarkably moving from 30 to 60 min (52 and 72 g/L, respectively), while it remains almost constant at longer reaction times.

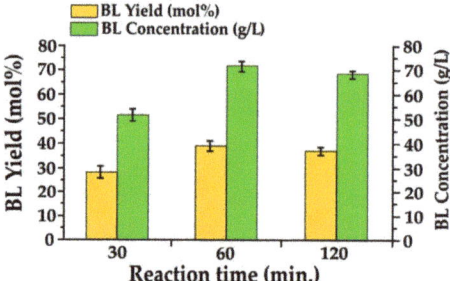

Figure 6. Kinetics of the one-pot butanolysis of the ADW *Eucalyptus nitens* sample (20 wt%) to BL. Reaction conditions: biomass loading 20 wt%, 190 °C, H_2SO_4 1.2 wt%, TR heating.

The increase of the final BL concentration is certainly a key parameter for the industrial scale-up of the reaction. In our case, only by acting on the biomass loading (up to the maximum of 20 wt%), it was possible to significantly increase the final BL concentration from the perspective of the high gravity approach. To further boost the concentration of the desired BL, it is possible to reprocess the mother liquor with fresh biomass, without adding further solvent and catalyst, according to the cross-flow approach [42]. This approach is certainly advantageous and smart, especially starting from cheap biomasses, as in our case. In this regard, the mother liquor deriving from the best alcoholysis run carried out under TR heating (run at 60 min, Figure 6) was used for a subsequent analogous alcoholysis reaction. In this additional step, a lower biomass loading (10 wt%) was adopted due to the very high viscosity of the liquor, and the results are reported in Table 3.

The final high BL concentration achieved with an additional alcoholysis step (85 g/L) justifies the validity of our approach. Regarding the BL molar yield, the small decrease occurred in the second step, despite the lower biomass loading, is probably due to the presence of reaction by-products obtained at the end of the first alcoholysis step, such as dibutyl ether (DBE), in other words to the lower amount of

n-BuOH available for the second alcoholysis step. However, these differences, in terms of BL molar yield, are not significant, being largely rewarded by the increase in BL concentration, the latter being a much more important process output, especially from an industrial perspective, allowing significant cost reduction of purification and separation treatments.

Table 3. Cross-flow butanolysis of ADW *Eucalyptus nitens* to BL. Reaction conditions: 190 °C, 60 min, H_2SO_4 1.2 wt% (added only at the 1st step), TR heating.

Step	Biomass Loading (wt%)	BL Yield (mol %)	BL Conc. (g/L)
1st	20	39	72
2nd	10	21	85

2.3. Design of Experiments and Optimization by RSM

The above promising preliminary results, prompted us to study the combined effect of three main factors, including temperature, reaction time and catalyst loading, on the butanolysis reaction, adopting an FCCD ($\alpha = 1$). BL molar yield was chosen as the response of interest, but the other main components of the reaction mixture were also determined. The ranges of the independent variables for planning the DOE were selected on the previous OFAT screening: temperature, x_1 (160–200 °C); reaction time, x_2 (30–180 min), catalyst loading, x_3 (0.2–3 wt%). These actual parameters were coded in three levels according to Equation (1):

$$X_i = (x_i - x_0)/\Delta x \tag{1}$$

where X_i is the coded value of the independent variable, x_i is the real value of the independent variable, x_0 is the real value of the independent variable at the center point, and Δx is the step change value. The complete case studies of 18 experiments, realized at the constant biomass loading of 20 wt%, are shown in Table 4, together with the respective experimental responses.

Table 4. Experimental design and BL molar yield (%) response of the Face-Centered Central Composite Design (FCCD) for different combinations of temperature, reaction time, and catalyst loading, all realized at the constant biomass loading of 20 wt%.

Run	Coded Parameter (Temp.) X_1	Coded Parameter (Time) X_2	Coded Parameter (Cat. Loading) X_3	Actual Parameter (Temp.) x_1, °C	Actual Parameter (Time) x_2, min	Actual Parameter (Cat. Loading) x_3, wt%	BL Molar Yield Y, mol%
1	1	1	−1	200	180	0.2	38
2	−1	0	0	160	105	1.6	35
3	−1	1	−1	160	180	0.2	6
4	1	−1	−1	200	30	0.2	20
5	−1	1	1	160	180	3.0	42
6	0	−1	0	180	30	1.6	36
7	0	0	1	180	105	3.0	40
8	1	1	1	200	180	3.0	34
9	0	0	0	180	105	1.6	40
10	1	0	0	200	105	1.6	37
11	1	1	−1	200	180	0.2	33
12	1	−1	1	200	30	3.0	41
13	0	0	0	180	105	1.6	42
14	−1	−1	−1	160	30	0.2	0
15	−1	−1	1	160	30	3.0	25
16	0	0	0	180	105	1.6	42
17	0	1	0	180	180	1.6	41
18	0	0	−1	180	105	0.2	21

The experimental data were analyzed by Design-Expert software and a second-order polynomial model was developed to correlate the process parameters with the response, thus obtaining Equation (2):

$$Y = 40.54 + 6.27X_1 + 3.97X_2 + 9.63X_3 - 1.41X_1X_2 - 5.59X_1X_3 - 1.84X_2X_3 - 3.95X_1^2 - 1.45X_2^2 - 9.45X_3^2 \tag{2}$$

According to the monomial coefficient value of the regression model equation, the order of priority among the main effects of impact factors is the following: catalyst loading > temperature > reaction time. Linear parameters have a significant synergistic effect on the response, since they have a positive coefficient, whereas the remaining combined and quadratic terms show significant antagonistic effects, thus highlighting the importance of the DOE optimization.

In Table 5, the results of the analysis of variance (ANOVA) are summarized to test the soundness and suitability of the model [59]. The mean squares values were calculated by dividing the sum of the squares of each variation source by their degrees of freedom, and a 95% confidence level was used to determine the statistical significance in all analyses. Results were assessed with p-value and F-value as the main statistical parameters of interest. The R^2 value for the quadratic model is 0.9102, which demonstrates a close agreement between experimental and predicted values of the BL molar yield. The R^2 adjusted is 0.8091 (>0.6), expressing that the model is significant. An adequate precision value of 11.2385 (>4) highlights a good signal, therefore this model can be used to describe the design space. Regarding the other parameters, both high F- and low p-values indicate the high significance of the corresponding coefficients of the model [58]. In our case, the model has an F-value of 9.01 (much greater than unity) and a p-value of 0.0025 (<0.05), which also implies that the model is significant. There is only a 0.25% chance that such a large F-value could occur due to noise. F-values of C, A, AC, and C^2 in Table 5 show that these are significant model terms, firstly C (catalyst loading) and secondly A (temperature), thus confirming the key roles of both these reaction variables on the butanolysis reaction, while reaction time has a modest effect. This latter aspect is certainly assessed to the chemistry of the butanolysis reaction, but also to the very efficient MW heating, which narrows the range of reaction time for reaching the optimal BL molar yield. Moreover, the same order of importance of the independent variables is deduced taking into account the p-values, whose significances are ascertained because of $p < 0.05$.

Table 5. ANOVA for the response surface quadratic model.

Source	Sum of Squares	Degree of Freedom	Mean Squares	F-Value	p-Value	Remark
Model	2452.42	9	272.49	9.01	0.0025	Significant
A—Temp.	416.24	1	416.24	13.76	0.0060	
B—Time	166.85	1	166.85	5.51	0.0468	
C—Cat. Load.	982.23	1	982.23	32.46	0.0005	
AB	17.18	1	17.18	0.5676	0.4728	
AC	268.38	1	268.38	8.87	0.0177	
BC	29.01	1	29.01	0.9586	0.3562	
A^2	41.99	1	41.99	1.39	0.2726	
B^2	5.67	1	5.67	0.1873	0.6766	
C^2	240.22	1	240.22	7.94	0.0226	
Residual	242.08	8	30.26			
Lack of fit	226.91	5	45.38	8.98	0.0503	Not significant
Pure error	15.17	3	5.06			
Cor Total						
$R^2 = 0.9102$	2694.50	17				
$R^2_{adj} = 0.8091$						

$p < 0.05$ is considered significant.

Diagnostic plots (predicted vs. actual plot and normal plot of residues) were checked for the adequacy and accuracy of the proposed model equation. The predicted vs. actual plot indicates that the points should be aligned with a straight line, and the normal plot of residues shows whether the residuals are in normal distribution [60]. A predicted vs. an actual plot of BL molar yield is shown in Figure 7a, which shows as the predicted values are close to the observed ones, in agreement with the above discussion. Additionally, the residuals showed a good fit to a normal distribution, indicating a high significance (Figure 7b).

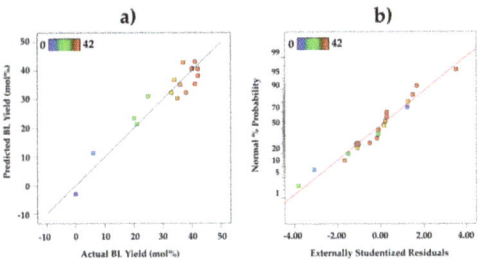

Figure 7. (**a**) Predicted versus actual plot, and (**b**) normal plot of residues.

Design-Expert software was used to produce three-dimensional (3D) response surfaces and two-dimensional (2D) contour plots. The 3D surfaces and 2D contour plots are graphical representations of the regression equation for the optimization of reaction conditions, which are very useful to visualize the relationship between the response variables and experimental levels of each factor. In such plots, the response functions of two factors are presented, while the remaining factor is kept constant at the central values. These graphs are shown in Figure 8.

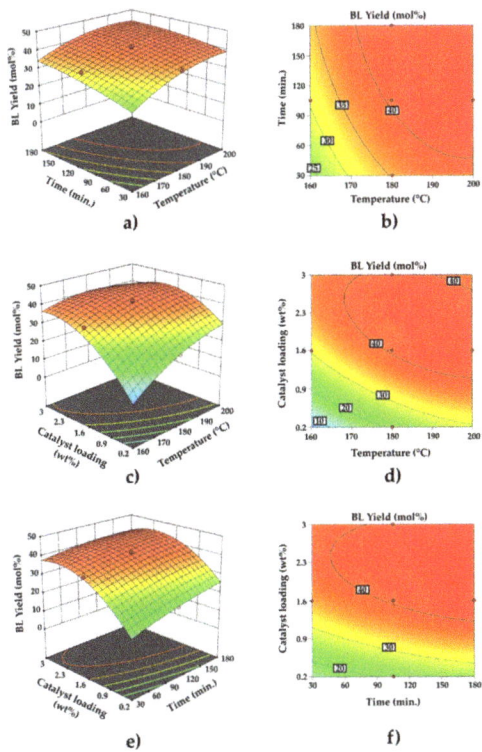

Figure 8. Three-dimensional (3D) response surfaces and two-dimensional (2D) contour plots: effect of temperature, reaction time, and acid concentration on BL molar yield. (**a,b**) catalyst loading was kept constant at 1.6 wt%; (**c,d**) reaction time was kept constant at 105 min; (**e,f**) temperature was kept constant at 180 °C.

Figure 8a,b confirms that the positive effect of the temperature on BL molar yield is more significant than that of the reaction time, at constant catalyst loading. In detail, a temperature higher than 180 °C is necessary to ensure the highest BL molar yield (about 40 mol%), together with relatively short times (up to about 120 min). On the other hand, a temperature lower than 180 °C is insufficient to achieve high BL molar yield, regardless of the adopted reaction times. Figure 8c,d elucidates that, firstly the catalyst loading, and secondly the temperature, strongly affect the BL yield, at constant time. This trend provides that a catalyst loading higher than 1.6 wt% is necessary to accomplish a very high BL molar yield, and that a lower catalyst loading should be associated with a corresponding higher reaction temperature (190–200 °C). Lastly, Figure 8e,f further confirms that, at a constant temperature, the catalyst loading has a strong non-linear influence on BL molar yield, and it should be higher than 1.6 wt%, while reaction time has a weaker effect, showing a feeble and presumably not significant curvature, as previously stated.

Starting from the above discussion, it is evident that the optimal solution for the BL optimization is not univocal, but involves rather a spatial region of the 3D response surfaces and 2D contour plots, depending on the combined choice of independent variables. The final stage of the design is the determination of the criteria for optimization and model validation. The optimization criterion was the maximum BL molar yield within the space design, with the independent variables kept within the range. For this purpose, starting from the acquired response surfaces and contour plots (Figure 8), the ranges of catalyst loading, temperature and reaction time, were further narrowed to those of greatest and practical interest for maximizing BL molar yield, avoiding the highest levels, in agreement with a more sustainable optimization approach. The identified ranges of interest were selected as follows: 180–190 °C for the temperature, 1.6–2.3 wt% for the catalyst loading, and 90–150 min for the reaction time. One of the possible solutions at the optimum levels (183 °C, 146 min, H_2SO_4 1.9 wt%) was experimentally carried out, and the experimental BL molar yield was compared with that predicted, as shown in Table 6. The results confirm the good agreement between the predictive and experimental results, at the optimum levels for BL synthesis, thus demonstrating the validity of our proposed model.

Table 6. Predicted and experimental BL yield: model validation.

Run	Actual Parameter x_1, °C	Actual Parameter x_2, min	Actual Parameter x_3, wt%	BL Yield (mol%) Predicted	BL Yield (mol%) Experimental	Desirability
19	183	146	1.9	44	42	1000

2.4. Identification of the Reaction By-Products and Application Perspectives of the Final Reaction Mixture

Before developing the possible engine applications of the final alcoholic mixture, it is necessary to analyze more in-depth its chemical composition, to better define its final use as biofuel. In this context, some authors have identified some reaction intermediates/by-products, but their quantification has not been reported [37], which is very useful for dealing with an in-depth discussion about the possible applications of this alcoholic mixture. Some possible intermediates/by-products were already defined in the Section 1 (Figure 2), in particular, furanic derivatives and glucosides as main reaction intermediates, and BF as the main reaction co-product. Furanic intermediates are very reactive species, which could condense to give solid polyfurans, or *humins* [40] and, in our case, their partial solubilization in the alcoholic mixture is more favored, if compared to the traditional hydrothermal path due to the presence of the alcoholic solvent, which acts as a polymerization inhibitor for *humins* growth [21,61]. Taking into account the chemical composition of ADW *Eucalyptus nitens*, which has a significant content of acetyl groups (4.8 wt%), deriving from the upstream Acetosolv treatment, these groups can be released during the alcoholysis, thus enabling the acid-catalyzed formation of butyl acetate (BA). Lastly, *n*-BuOH can be etherified to give dibutyl ether (DBE) and water, the latter in equimolar amount respect to DBE, and also this reaction favorably occurs in the presence of the adopted sulfuric acid catalyst [37].

To confirm the presence of the above by-products, the reaction mixtures recovered from the experiments planned for the FCCD (Table 4) and the model validation (Table 6), were qualitatively analyzed by GC–MS, identifying BF, DBE, and BA as main reaction by-products, together with the unconverted n-BuOH. These compounds were subsequently quantified by GC–FID, and the corresponding mass yields in the organic phase are reported in Table 7, together with that of the product of interest (BL).

Table 7. Composition of the organic reaction mixtures recovered from the experiments planned for the FCCD and for the model validation, working at the constant biomass loading of 20 wt%.

Run	Temperature (°C)	Time (min)	Catalyst Loading (wt%)	Composition (wt%)			
				BL	DBE	BA	n-BuOH
1	200	180	0.2	9	18	3	70
2	160	105	1.6	9	13	3	76
3	160	180	0.2	1	4	3	91
4	200	30	0.2	6	6	3	85
5	160	180	3.0	11	23	3	63
6	180	30	1.6	9	11.7	3	77
7	180	105	3.0	10	37.9	3	49
8	200	180	3.0	10	58.3	3	28
9	180	105	1.6	10	26.8	3	61
10	200	105	1.6	9	45.0	3	43
11	200	180	0.2	7	10.6	3	80
12	200	30	3.0	10	37.9	3	49
13	180	105	1.6	10	23.8	3	63
14	160	30	0.2	0	0.8	2	97
15	160	30	3.0	7	9.4	3	81
16	180	105	1.6	10	26.2	3	61
17	180	180	1.6	10	33.3	3	54
18	180	105	0.2	5	7.1	3	85
19	183	146	1.9	11	39.4	3	47

The above data show that the variation of BA and BL yield within the investigated ranges of the independent variables is modest, if compared with that of DBE (and consequently that of n-BuOH), which represents the main reaction by-product, even in the case of the optimum experiment for BL synthesis (Run 19, Table 7). DBE represents a high cetane component (CN = 100) and it was already tested in blend with diesel fuel, leading to very short ignition delays, so its possible application in compression ignition engines is favorable and attractive [22]. However, different experimental conditions should allow a significant modulation of the DBE to n-BuOH weight ratio, etherification being significantly favored by the acidity increase (compare Runs 1 and 8). On this basis, the best experimental choice for performing the biomass butanolysis should lead to a good production of both BL and DBE, while the unconverted n-BuOH could be eventually recovered and reused within the same process [62]. In principle, the organic ternary mixture BL/DBE/n-BuOH could be immediately exploited, without separation of its components as an innovative diesel fuel additive, thus making the alcoholysis reaction a viable route to the direct production of a blending component. In addition, the amount of the adopted mineral acid for the butanolysis reaction should be as low as possible, to avoid costly work-up procedures and, on this basis, the reaction mixture deriving from Run 1 represents the best compromise for developing the next application of this mixture as a diesel additive. In order to explore this never reported perspective, a preliminary study was carried out employing a model mixture BL/DBE/n-BuOH as an additive for diesel fuel. At this preliminary level of investigation, the addition of BA, which is a minor product closely related to the adopted biomass, was not considered.

2.5. Engine Experimental Activity

A preliminary engine experimental activity was carried out to verify the influence of these oxygenated fuel additives on diesel engine performance. As aforementioned, the ternary mixture available from one-pot butanolysis of raw and ADW biomass is mainly composed of BL, DBE, and unreacted n-BuOH. These compounds represent valuable oxygenated fuels and their properties were already investigated, singularly, in blend with diesel [22,63–66]. The properties of all fuel components are shown in Table 8 and compared with those of commercial diesel fuel.

Table 8. Overview of biomass resources available from the literature [13,21,61].

Properties	DBE	BL	n-BuOH	Diesel
$T_{evaporation}$ (°C)	140	232	118	180–360
O_2 (wt%)	12	28	21.6	0
Density (g/L)	769 [a]	974 [a]	810 [b]	837 [b]
CN [c] (−)	100	14	25	50
LHV [d] (MJ/Kg)	42.8	27.4	33.1	43
ν [e] (mm^2/s)	0.72	1.5	2.22	2.6

[a] Density at 25 °C. [b] Density at 20 °C. [c] Cetane Number. [d] Lower Heating Value. [e] Kinematic viscosity at 40 °C.

The above properties show that both n-BuOH and BL have a lower Cetane Number than diesel. This is a known behavior reported by Koivisto et al. [66,67] for alcohols and levulinates, including n-BuOH and BL, respectively. These compounds are characterized by higher ignition delays (i.e., lower Cetane Number) than alkanes of the same carbon atom chain length. However, ethers, such as DBE, have an opposite behavior and show lower ignition delays in comparison with alkanes [67]. Taking into account the components of the ternary mixture obtained from alcoholysis reaction of biomass, DBE can play a fundamental role as a cetane enhancer, making its controlled coproduction in the alcoholysis highly valuable in this applicative perspective. Moreover, this characteristic enables us to test diesel blended with a high-volume percentage of the ternary mixture. In addition, the use of the n-BuOH/DBE/BL mixture leads to an increase in the fuel oxygen content. Generally, oxygenated diesel blends ensure, especially in the areas of the cylinder with a low air-to-fuel ratio, the presence of oxygen directly from the fuel and, consequently, soot precursors reduction [22,68]. In this context, n-BuOH and BL represent the two components of the mixture which mostly influence the oxygen content in the final blend with diesel.

By considering that the final composition of the reaction can be easily tuned, a model mixture with a composition similar to that of Run 1 in Table 7 was prepared and tested on a small diesel engine. The mixture, whose composition is reported in Table 9, was blended in three different volume percentages with diesel: 10, 20, and 30 vol % (named MIX1 10%, MIX1 20%, MIX1 30%).

Table 9. Composition of the prepared ternary mixture.

Mixture	n-BuOH (wt%)	DBE (wt%)	BL (wt%)
MIX1	70	20	10

This mixture was utilized in the experimental engine at different rpm values (1500, 2000, and 2500 rpm) at full power (T_{max}). Data of the engine performance were compared with those obtained with diesel fuel alone, and all the data of the engine performances are reported in Figure 9. Fuel injection timing was maintained constant along with the experimentation.

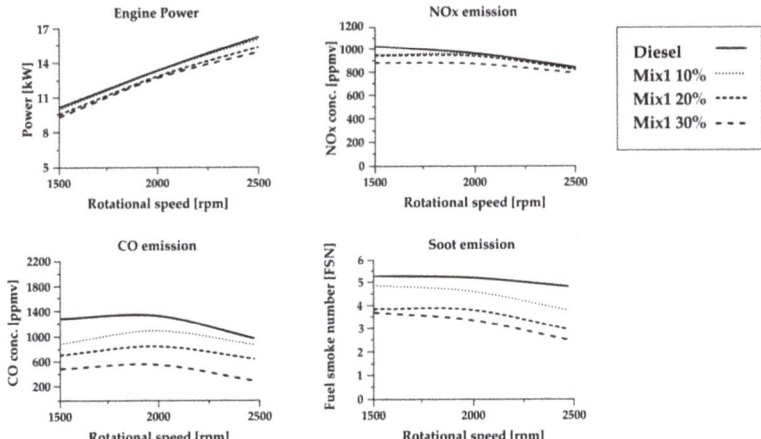

Figure 9. Engine performances obtained with diesel fuel, alone and in blends with *n*-BuOH/dibutyl ether (DBE)/BL.

The use of different blending (from 10 to 30 vol %) with diesel fuel does not significantly affect the power of the employed engine, meaning that the calorific values and the reactivity of these mixtures do not significantly differ from those of commercial diesel fuel. Moreover, no considerable variation of HC and NOx emissions occurs with the use of the adopted mixtures. Literature reports different nitrogen oxides behavior as a function of the fuel type, which indicates that its emission strongly depends on several other factors besides the employed fuel, and these are not easily detachable within the performed analysis [69,70].

On the other hand, a strong reduction of both CO and soot emission was obtained as the blend of both the ternary mixtures was increased. This can be addressed to the increased combustion oxygen availability which plays a key role in the formation process of the carbon-based pollutants, such as soot or CO. Particularly, the oxygen provided by the fuel reduces the low air excess zones which are the main cause of the soot formation. Furthermore, more oxygen is available from fuel, and less carbon is present for CO or soot formation. Svensson et al. [71] found that the soot emissions could be reduced to 0 when fuel oxygen content reaches 27–35 wt%. Another aspect that contributes to decreased CO and soot when applying the two mixtures is the lower boiling point of the oxygenated components in comparison with diesel fuel. This leads to a kind of "droplet explosion" once the fuel mixture is introduced in a hot ambient, such as the cylinder at the end of the compression stroke, increasing the spray fragmentation and mixing [72,73], so increasing the combustion completion of carbon-based molecules is necessary.

3. Materials and Methods

3.1. Materials

Eucalyptus nitens was collected locally in the Galicia region (Spain). The starting untreated and ADW *Eucalyptus nitens* biomasses were milled with a knife mill, using a 0.5 mm metal mesh, air-dried, and further processed according to the following pre-treatments. Regarding the ADW treatment, the raw *Eucalyptus nitens* was first subjected to autohydrolysis. For this purpose, the sample was suspended in water and treated in a stainless steel reactor (Parr Instruments Company, Moline, IL, USA) under non-isothermal conditions, adopting a water/biomass weight ratio of 10/1, up to the final temperature of 193 °C [51]. Then, the solid residue was recovered and subsequently underwent an Acetosolv treatment, with a mixture of 90.00% acetic acid, 9.78% water, and 0.22% hydrochloric acid. For this purpose, the temperature was maintained at 134 °C for 30 min, adopting the solid/liquid weight ratio of 1/10 [51]. Solid residue obtained from this treatment was recovered by filtration, washed with water, and finally air-dried.

3.2. Characterization of the Starting Eucalyptus nitens Samples

The compositional analysis of the starting untreated and ADW *Eucalyptus nitens* samples was carried out based on the standard NREL procedures [74,75]. XRD analysis was carried out using a vertical goniometer diffractometer D2-PHASER (Bruker, Billerica, MA, USA). The analyses were performed using the CuKα radiation at 1.54 Å as the X-ray source. The interval used was $5° < 2θ < 40°$, with a resolution of 0.016°. DIFFRAC software (Bruker AXS, Karlsruhe, Germany) was used for spectra processing. The crystallinity index of the *Eucalyptus nitens* samples was calculated after deconvolution of the curves, which was carried out by the PeakFit software (Systat Software Inc., San Jose, CA, USA), taking into account the contribution of the amorphous component (at about $2θ = 21.5°$) and the peaks related to the crystalline plans with Miller indices 101, 10ī, 002 e 040, as reported in the literature [52]. The integration of the areas of these peaks allowed the estimation of the crystallinity index (CI) of the cellulose, based on Equation (3):

$$CI(\%) = [1 - (A_{AM}/A_{Total})] \times 100 \qquad (3)$$

where A_{AM} is the area of the peak corresponding to the amorphous cellulose, and A_{TOT} is the total area of all peaks.

Fourier Transform–Infrared (FT–IR) characterization of the biomass samples was performed with a Perkin-Elmer Spectrum-Two spectrophotometer (Perkin–Elmer, Waltham, MA, USA), equipped with an Attenuated Total Reflectance (ATR) apparatus. The acquisition of each spectrum has provided 12 scans, with a resolution of 8 cm^{-1}, in the wavenumber range between 4000 and 450 cm^{-1}.

3.3. Alcoholysis Experiments

MW-assisted alcoholysis of the untreated and ADW *Eucalyptus nitens* samples to BL was performed in the single-mode MW reactor (CEM Discover S-class System), employing the 35 mL vessel with a Teflon stir bar. Once the starting biomass, *n*-BuOH, *n*-dodecane (internal standard) and sulfuric acid (catalyst) were weighed in the vessel, the reactor was closed and the sealed system was irradiated up to the set-point temperature. The maximum pulsed-power of 300 W was used to heat the samples. During the reaction, pressure and temperature values were continuously acquired with the software and controlled with a feedback algorithm to maintain the constant temperature. At the end of each hydrolysis reaction, the reactor was rapidly cooled at room temperature by blown air and the solid-liquid slurry was recovered, filtered under vacuum, properly diluted with acetone, and analyzed by Gas Chromatography.

Alcoholysis experiments with conventional heating were carried out in the 60 mL glass reactor. Once the starting biomass, *n*-BuOH, *n*-dodecane (internal standard) and sulfuric acid (catalyst) were weighed in the reactor, it was closed and the sealed system was placed in an oil bath, previously heated to the set-point temperature. At the end of each alcoholysis reaction, the reactor was rapidly cooled at room temperature by blown air and the solid-liquid slurry was recovered, filtered under vacuum, properly diluted with acetone, and analyzed by Gas Chromatography. For this purpose, the reaction products were qualitatively identified by Gas Chromatography coupled with Mass Spectrometry (GC–MS), and subsequently quantified by Gas Chromatography coupled with Flame Ionization Detector (GC–FID). Regarding GC–MS analysis, a gas chromatograph Hewlett-Packard (Hewlett-Packard HP, Palo Alto, CA, USA) HP 6890 equipped with an MSDHP 5973 detector and with a G.C. column Phenomenex Zebron with a 100% methyl polysiloxane stationary phase (30 m × 0.25 mm × 0.25 µm), was used. The transport gas was helium 5.5 and the flow was 1 mL/min. The temperatures of the injection port and detector were set at 250 °C and 290 °C, respectively. The carrier pressure at 100 kPa and the split flow at 3.40 ms^{-1} were adopted. The oven was heated at 60 °C for 3 min, and then the temperature was raised at 10 °C/min up to 260 °C for 5 min, and lastly, 10 °C/min up to 280 °C for 3 min. GC–FID analysis was carried out by a DANI GC 1000 DPC (Dani Instruments S.P.A., Cologno Monzese, Italy) gas chromatograph, equipped with a fused silica capillary column—HP-PONA cross-linked methyl silicone gum (20 m × 0.2 mm × 0.5 µm). The FID ports were set at 250 °C. The oven temperature program was set at 90 °C for 3 min and then increased at the rate of 10 °C/min up to 260 °C, where it was maintained for 5 min, then up to 280 °C with the rate of 10 °C/min and maintained for 3 min. Nitrogen was used as the carrier gas, at the flow rate of 0.2 mL/min. Quantitative determination of BL, DBE, BA, and unconverted n-BuOH was carried out with the internal standard method, using n-dodecane as the internal standard. Each analysis was carried out in duplicate and the reproducibility of the technique was within 5%.

The yield to BL was calculated as follows:

$$\text{Yield to BL (mol\%)} = (\text{mol BL/mol } C_6H_{10}O_5 \text{ units in the starting biomass}) \times 100 \qquad (4)$$

Besides, in the case of the Cross-Flow experiments, BL yield was calculated as follows:

$$\text{Yield to BL (mol\%)} = (\text{mol BL obtained in the 2nd step/mol } C_6H_{10}O_5 \text{ units in the biomass added in the 2nd step}) \times 100 \qquad (5)$$

3.4. Experimental Design

Response Surface Methodology (RSM) and Face-Centered Central Design (FCCD) were employed for the reaction optimization by maximizing the response, that is BL molar yield, investigating appropriate ranges of the independent variables. The chosen independent variables are temperature, reaction time, and catalyst loading, as reported in Table 4. Their levels were selected starting from preliminary One-Factor-at-a-Time (OFAT) experiments. The experimental design in this study required 18 experimental runs, which included 4 replicates. The software Design-Expert 12 (12.0.1.0) Trial Version (Stat-Ease, Inc., Minneapolis, MN, USA) was adopted to process and analyze the results. The data were fitted to the polynomial model and presented in the analysis of variance (ANOVA). The quadratic equation that represents the correlation between independent variables and the response can be expressed by the quadratic polynomial Equation (6):

$$Y = b_0 + b_1X_1 + b_2X_2 + b_3X_3 + b_{11}X_1^2 + b_{22}X_2^2 + b_{33}X_3^2 + b_{12}X_1X_2 + b_{13}X_1X_3 + b_{23}X_2X_3 \qquad (6)$$

where Y is the predicted response, b_0 the constant, b_1, b_2, and b_3 the linear coefficients, b_{12}, b_{13}, and b_{23} the cross-product coefficients, and b_{11}, b_{22}, and b_{33} are the quadratic coefficients.

3.5. Engine Experimental Setup

A small diesel engine, whose specifications are reported in Table 10, was chosen and coupled with a Borghi and Saveri eddy current brake with rpm/torque controller. An AVL gravimetric fuel balance was used to online measure fuel consumption. An Environnement SA test bench, equipped with a Non-Dispersive Infra-Red (NDIR) Sensor, a paramagnetic sensor, a Heated Chemiluminescence Detector (HCLD), and a Heated Flame Ionization Detector (HFID) was employed to measure, respectively, CO and CO_2, O_2, NOx, and THC (Total Hydro-Carbons). The particulate matter was determined using a dedicated sample line and an AVL smoke meter. An exhaust gas K-type thermocouple was employed to verify the occurrence of the steady-state conditions, for each different test condition. Once the engine was stabilized in a particular operating condition, data were collected and analyzed to provide average values.

Table 10. Experimental engine characteristics.

Engine Type	Lombardini LD 625/2
Number of cylinders	2
Cooling system	Forced air
Displacement (cm^3)	1248
Bore (mm)	95
Stroke (mm)	88
Compression ratio	17.5:1
Max rotational speed (rpm)	3000
Power @ 3000 rpm (kW)	21
Max Torque @ 2200 rpm (Nm)	29.4
Fuel injection system	Direct Mechanic

4. Conclusions

In this work, autohydrolyzed-delignified *Eucalyptus nitens* wood was employed as cheap cellulose-rich feedstock for the one-pot alcoholysis to *n*-butyl levulinate, adopting *n*-butanol as green reagent/reaction medium, very dilute sulfuric acid as a homogeneous catalyst, and microwave as an efficient heating system. The effect of the main reaction parameters to *n*-butyl levulinate was investigated firstly by a traditional One-Factor-at-a-Time approach, to verify the feasibility of this reaction and identify the coarse ranges of the operating variables. Under the best reaction conditions (microwave heating, 190 °C, 15 min, biomass loading 20 wt%, 1.2 wt% H_2SO_4), the maximum *n*-butyl levulinate molar yield of about 30 mol% was achieved, using a very high biomass loading (20 wt%), an eligible aspect from the perspective of an intensified *high gravity* approach. However, even higher molar yields (up to about 40 mol%) were obtained adopting traditional heating (190 °C, 120 min, biomass loading 20 wt%, 1.2 wt% H_2SO_4), demonstrating the good feasibility of the reaction also with traditional heating systems, aspect of paramount industrial interest. The possibility of reprocessing the reaction mixture deriving from the optimized experiment by addition of fresh biomass, was evaluated, achieving the maximum *n*-butyl levulinate concentration of about 85 g/L after only one reprocessing of the mother liquor, and this is the highest *n*-butyl levulinate concentration hitherto reported in the literature starting from real biomass.

The butanolysis reaction was further optimized by Response Surface Methodology, utilizing a Face-Centered Central Composite Design. The chosen design appropriately describes the studied real system, which requires mild acidity and high temperature, for maximizing *n*-butyl levulinate production, while the effect of the reaction time is softened due to the efficient microwave heating. The significance of the independent variables and their possible interactions was tested using ANOVA with a 95% confidence level, and the model was experimentally validated at the optimal operating conditions for *n*-butyl levulinate production.

Finally, a preliminary study of diesel engine performances and emissions for a model mixture with a composition analogous to that of the main components of the reaction mixture was performed, to draw an indication of its potential application as an additive for diesel fuel, without performing the separation of each component.

Author Contributions: S.G., A.M.R.G., and S.F. conceived the experiments; S.G., A.M.R.G, C.A., S.F., M.L., D.L., and G.P. designed the experiments; S.G., M.L., and G.P. performed the experiments and analysis; all the authors analyzed the data; D.L. and S.G. wrote the paper; J.C.P., S.F., A.M.R.G., and C.A. revised and supervised the writing of the manuscript. All authors have read and agreed to the published version of the manuscript.

Funding: The Spanish Ministry of Economy and Competitivity supported this study in the framework of the research project "Modified aqueous media for wood Biorefineries" (reference CTQ2017-82962-R), partially funded by the FEDER program of the European Union. Mar López thanks the European Social Fund (ESF) for economic support and the "Xunta de Galicia" for her predoctoral grant (reference ED481A-2017/316). MIUR supported this study in the framework of the research project VISION PRIN 2017 FWC3WC_002.

Conflicts of Interest: The authors declare no conflict of interest.

References

1. Girisuta, B.; Heeres, H.J. Levulinic acid from biomass: Synthesis and applications. In *Production of Platform Chemicals from Sustainable Resources*; Fang, Z., Smith, R.L., Qi, X., Eds.; Springer Nature Pte Ltd.: Singapore, 2017; Volume 7, pp. 143–169. [CrossRef]
2. Available online: https://www.psmarketresearch.com/market-analysis/levulinic-acid-market (accessed on 23 March 2020).
3. Rivas, S.; Raspolli Galletti, A.M.; Antonetti, C.; Santos, V.; Parajó, J.C. Sustainable production of levulinic acid from the cellulosic fraction of *Pinus Pinaster* wood: Operation in aqueous media under microwave irradiation. *J. Wood Chem. Technol.* **2015**, *35*, 315–324. [CrossRef]
4. Antonetti, C.; Licursi, D.; Fulignati, S.; Valentini, G.; Raspolli Galletti, A.M. New frontiers in the catalytic synthesis of levulinic acid: From sugars to raw and waste biomass as starting feedstock. *Catalysts* **2016**, *6*, 196. [CrossRef]
5. Licursi, D.; Antonetti, C.; Martinelli, M.; Ribechini, E.; Zanaboni, M.; Raspolli Galletti, A.M. Monitoring/characterization of stickies contaminants coming from a papermaking plant—Toward an innovative exploitation of the screen rejects to levulinic acid. *Waste Manag.* **2016**, *49*, 469–482. [CrossRef] [PubMed]
6. Rivas, S.; Raspolli Galletti, A.M.; Antonetti, C.; Santos, V.; Parajó, J.C. Sustainable conversion of *Pinus Pinaster* wood into biofuel precursors: A biorefinery approach. *Fuel* **2016**, *164*, 51–58. [CrossRef]
7. Licursi, D.; Antonetti, C.; Mattonai, M.; Pérez-Armada, L.; Rivas, S.; Ribechini, E.; Raspolli Galletti, A.M. Multi-valorisation of giant reed (*Arundo Donax* L.) to give levulinic acid and valuable phenolic antioxidants. *Ind. Crop. Prod.* **2018**, *112*, 6–17. [CrossRef]
8. Deémolis, A.; Essayem, N.; Rataboul, F. Synthesis and applications of alkyl levulinates. *ACS Sustain. Chem. Eng.* **2014**, *2*, 1338–1352. [CrossRef]
9. Badgujar, K.C.; Badgujar, V.C.; Bhanage, B.M. A review on catalytic synthesis of energy rich fuel additive levulinate compounds from biomass derived levulinic acid. *Fuel Process. Technol.* **2020**, *197*, 106213. [CrossRef]
10. Chung, Y.-H.; Peng, T.-H.; Lee, H.-Y.; Chen, C.-L.; Chien, I.-L. Design and control of reactive distillation system for esterification of levulinic acid and *n*-butanol. *Ind. Eng. Chem. Res.* **2015**, *54*, 3341–3354. [CrossRef]
11. Li, C.; Duan, C.; Fang, J.; Li, H. Process intensification and energy saving of reactive distillation for production of ester compounds. *Chin. J. Chem. Eng.* **2019**, *27*, 1307–1323. [CrossRef]
12. Vázquez-Castillo, J.A.; Contreras-Zarazúa, G.; Segovia-Hernández, J.G.; Kiss, A.A. Optimally designed reactive distillation processes for eco-efficient production of ethyl levulinate. *J. Chem. Technol. Biotehnol.* **2019**, *94*, 2131–2140. [CrossRef]
13. Christensen, E.; Williams, A.; Paul, S.; Burton, S.; McCormick, R.L. Properties and performance of levulinate esters as diesel blend components. *Energy Fuels* **2011**, *25*, 5422–5428. [CrossRef]
14. Li, H.; Peng, L.; Lin, L.; Chen, K.; Zhang, H. Synthesis, isolation and characterization of methyl levulinate from cellulose catalyzed by extremely low concentration acid. *J. Energy Chem.* **2013**, *22*, 895–901. [CrossRef]
15. Ahmad, E.; Alam, M.I.; Pant, K.K.; Haider, M.A. Catalytic and mechanistic insights into the production of ethyl levulinate from biorenewable feedstocks. *Green Chem.* **2016**, *18*, 4804–4823. [CrossRef]
16. Negahdar, L.; Al-Shaal, M.G.; Holzhäuser, F.J.; Palkovits, R. Kinetic analysis of the catalytic hydrogenation of alkyl levulinates to γ-valerolactone. *Chem. Eng. Sci.* **2017**, *158*, 545–551. [CrossRef]
17. Filiciotto, L.; Balu, A.M.; Van der Waal, J.C.; Luque, R. Catalytic insights into the production of biomass-derived side products methyl levulinate, furfural and humins. *Catal. Today* **2018**, *302*, 2–15. [CrossRef]
18. Shrivastav, G.; Khan, T.S.; Agarwal, M.; Haider, M.A. Reformulation of gasoline to replace aromatics by biomass-derived alkyl levulinates. *ACS Sustain. Chem. Eng.* **2017**, *5*, 7118–7127. [CrossRef]
19. Zhao, T.; Zhang, Y.; Zhao, G.; Chen, X.; Han, L.; Xiao, W. Impact of biomass feedstock variability on acid-catalyzed alcoholysis performance. *Fuel Process. Technol.* **2018**, *180*, 14–22. [CrossRef]
20. Hu, X.; Li, C.-Z. Levulinic esters from the acid-catalysed reactions of sugars and alcohols as part of a bio-refinery. *Green Chem.* **2011**, *13*, 1676–1679. [CrossRef]
21. Hu, X.; Wu, L.; Wang, Y.; Mourant, D.; Lievens, C.; Gunawan, R.; Li, C.-Z. Mediating acid-catalyzed conversion of levoglucosan into platform chemicals with various solvents. *Green Chem.* **2012**, *14*, 3087–3098. [CrossRef]

22. Kremer, F.; Pischinger, S. Butyl ethers and levulinates. In *Biofuels from Lignocellulosic Biomass: Innovations Beyond Bioethanol*; Boot, M., Ed.; Wiley-VCH GmbH & Co. KGaA: Weinheim, Germany, 2016; pp. 87–104. [CrossRef]
23. Christensen, E.; Yanowitz, J.; Ratcliff, M.; McCormick, R.L. Renewable oxygenate blending effects on gasoline properties. *Energ. Fuel* **2011**, *25*, 4723–4733. [CrossRef]
24. Benito, P.; Vaccari, A.; Antonetti, C.; Licursi, D.; Schiarioli, N.; Rodriguez-Castellón, E.; Raspolli Galletti, A.M. Tunable copper-hydrotalcite derived mixed oxides for sustainable ethanol condensation to *n*-butanol in liquid phase. *J. Clean. Prod.* **2019**, *209*, 1614–1623. [CrossRef]
25. Mishra, D.K.; Kumar, S.; Shukla, R.S. Chapter 12—Furfuryl alcohol—A promising platform chemical. In *Biomass, Biofuels, Biochemicals. Recent Advances in Development of Platform Chemicals*; Saravanamurugan, S., Pandey, A., Li, H., Riisager, A., Eds.; Elsevier: Amsterdam, The Netherlands, 2020; pp. 323–353. [CrossRef]
26. Gitis, V.; Chung, S.-H.; Shiju, N.R. Conversion of furfuryl alcohol into butyl levulinate with graphite oxide and reduced graphite oxide. *Flat Chem.* **2018**, *10*, 39–44. [CrossRef]
27. Gupta, S.S.R.; Kantam, M.L. Catalytic conversion of furfuryl alcohol or levulinic acid into alkyl levulinates using a sulfonic acid-functionalized hafnium-based MOF. *Catal. Commun.* **2019**, *124*, 62–66. [CrossRef]
28. Bernal, H.G.; Oldani, C.; Funaioli, T.; Raspolli Galletti, A.M. AQUIVION® perfluorosulfonic acid resin for butyl levulinate production from furfuryl alcohol. *N. J. Chem.* **2019**, *43*, 14694–14700. [CrossRef]
29. Yu, X.; Peng, L.; Pu, Q.; Tao, R.; Gao, X.; He, L.; Zhang, J. Efficient valorization of biomass-derived furfuryl alcohol to butyl levulinate using a facile lignin-based carbonaceous acid. *Res. Chem. Intermed.* **2020**, *46*, 1469–1485. [CrossRef]
30. Iborra, M.; Tejero, J.; Fité, C.; Ramírez, E.; Cunill, F. Liquid-phase synthesis of butyl levulinate with simultaneous water removal catalyzed by acid ion exchange resins. *J. Ind. Eng. Chem.* **2019**, *78*, 222–231. [CrossRef]
31. Yang, J.; Li, G.; Zhang, L.; Zhang, S. Efficient production of *n*-butyl levulinate fuel additive from levulinic acid using amorphous carbon enriched with oxygenated groups. *Catalysts* **2018**, *8*, 14. [CrossRef]
32. Garves, K. Acid catalyzed degradation of cellulose in alcohols. *J. Wood Chem. Technol.* **1988**, *8*, 121–134. [CrossRef]
33. Hishikawa, Y.; Yamaguchi, M.; Kubo, S.; Yamada, T. Direct preparation of butyl levulinate by a single solvolysis process of cellulose. *J. Wood Sci.* **2013**, *59*, 179–182. [CrossRef]
34. Liu, Y.; Lin, L.; Liu, D.; Zhuang, J.; Pang, C. Conversion of biomass sugars to butyl levulinate over combined catalyst of solid acid and other acid. *Adv. Mater. Res.* **2014**, *955*, 779–784. [CrossRef]
35. Ma, H.; Long, J.-X.; Wang, F.-R.; Wang, L.-F.; Li, X.-H. Conversion of cellulose to butyl levulinate in bio-butanol medium catalyzed by acidic ionic liquids. *Acta Phys. Chim. Sin.* **2015**, *31*, 973–979. [CrossRef]
36. Yamada, T.; Yamaguchi, M.; Kubo, S.; Hishikawa, Y. Direct production of alkyl levulinates from cellulosic biomass by a single-step acidic solvolysis system at ambient atmospheric pressure. *BioResources* **2015**, *10*, 4961–4969. [CrossRef]
37. Démolis, A.; Eternot, M.; Essayem, N.; Rataboul, F. Influence of butanol isomers on the reactivity of cellulose towards the synthesis of butyl levulinates catalyzed by liquid and solid acid catalysts. *N. J. Chem.* **2016**, *40*, 3747–3754. [CrossRef]
38. Elumalai, S.; Agarwal, B.; Runge, T.M.; Sangwan, R.S. Integrated two-stage chemically processing of rice straw cellulose to butyl levulinate. *Carbohydr. Polym.* **2016**, *150*, 286–298. [CrossRef] [PubMed]
39. An, R.; Xu, G.; Chang, C.; Bai, J.; Fang, S. Efficient one-pot synthesis of *n*-butyl levulinate from carbohydrates catalyzed by $Fe_2(SO_4)_3$. *J. Energy Chem.* **2017**, *26*, 556–563. [CrossRef]
40. Deng, L.; Chang, C.; An, R.; Qi, X.; Xu, G. Metal sulfates-catalyzed butanolysis of cellulose: Butyl levulinate production and optimization. *Cellulose* **2017**, *24*, 5403–5415. [CrossRef]
41. Liang, C.; Wang, Y.; Hu, Y.; Wu, L.; Zhang, W. Study of a new process for the preparation of butyl levulinate from cellulose. *ACS Omega* **2019**, *4*, 9828–9834. [CrossRef]
42. Rivas, S.; Raspolli Galletti, A.M.; Antonetti, C.; Licursi, D.; Santos, V.; Parajó, J.C. A biorefinery cascade conversion of hemicellulose-free *Eucalyptus Globulus* wood: Production of concentrated levulinic acid solutions for γ-valerolactone sustainable preparation. *Catalysts* **2018**, *8*, 169. [CrossRef]
43. Pérez-Cruzado, C.; Merino, A.; Rodríguez-Soalleiro, R. A management tool for estimating bioenergy production and carbon sequestration in *Eucalyptus globulus* and *Eucalyptus nitens* grown as short rotation woody crops in north-west Spain. *Biomass Bioenerg.* **2011**, *35*, 2839–2851. [CrossRef]

44. Peleteiro, S.; Raspolli Galletti, A.M.; Antonetti, C.; Santos, V.; Parajó, J.C. Manufacture of furfural from xylan-containing biomass by acidic processing of hemicellulose-derived saccharides in biphasic media using microwave heating. *J. Wood Chem. Technol.* **2018**, *38*, 198–213. [CrossRef]
45. Chen, X.; Zhang, K.; Xiao, L.-P.; Sun, R.-C.; Song, G. Total utilization of lignin and carbohydrates in *Eucalyptus grandis*: An integrated biorefinery strategy towards phenolics, levulinic acid, and furfural. *Biotechnol. Biofuels* **2020**, *13*, 1–10. [CrossRef] [PubMed]
46. Antonetti, C.; Licursi, D.; Raspolli Galletti, A.M.; Martinelli, M.; Tellini, F.; Valentini, G.; Gambineri, F. Application of microwave irradiation for the removal of polychlorinated biphenyls from siloxane transformer and hydrocarbon engine oils. *Chemosphere* **2016**, *159*, 72–79. [CrossRef] [PubMed]
47. Di Fidio, N.; Raspolli Galletti, A.M.; Fulignati, S.; Licursi, D.; Liuzzi, F.; De Bari, I.; Antonetti, C. Multi-step exploitation of raw *Arundo donax* L. for the selective synthesis of second-generation sugars by chemical and biological route. *Catalysts* **2020**, *10*, 79. [CrossRef]
48. Ahmad, E.; Alam, E.I.; Pant, K.K.; Haider, M.A. Insights into the synthesis of ethyl levulinate under microwave and non-microwave heating conditions. *Ind. Eng. Chem. Res.* **2019**, *58*, 16055–16064. [CrossRef]
49. Dai, J.; Peng, L.; Li, H. Intensified ethyl levulinate production from cellulose using a combination of low loading H_2SO_4 and $Al(OTf)_3$. *Catal. Commun.* **2018**, *103*, 116–119. [CrossRef]
50. Grisel, R.J.H.; van der Waal, J.C.; de Jong, E.; Huijgen, W.J.J. Acid catalysed alcoholysis of wheat straw: Towards second generation furan-derivatives. *Catal. Today* **2014**, *223*, 3–10. [CrossRef]
51. Penín, L.; Peleteiro, S.; Santos, V.; Alonso, J.L.; Parajo, J.C. Selective fractionation and enzymatic hydrolysis of *Eucalyptus nitens* wood. *Cellulose* **2019**, *26*, 1125–1139. [CrossRef]
52. Park, S.; Baker, J.O.; Himmel, M.E.; Parilla, P.A.; Johnson, D.K. Cellulose crystallinity index: Measurement techniques and their impact on interpreting cellulase performance. *Biotechnol. Biofuels* **2010**, *3*, 1–10. [CrossRef]
53. Banerjee, D.; Mukherjee, S.; Pal, S.; Khowala, S. Enhanced saccharification efficiency of lignocellulosic biomass of mustard stalk and straw by salt pretreatment. *Ind. Crops Prod.* **2016**, *80*, 42–49. [CrossRef]
54. Fan, S.; Zhang, P.; Li, F.; Jin, S.; Wang, S.; Zhou, S. A review of lignocellulose change during hydrothermal pretreatment for bioenergy production. *Curr. Org. Chem.* **2016**, *20*, 1–11. [CrossRef]
55. Licursi, D.; Antonetti, C.; Bernardini, J.; Cinelli, P.; Coltelli, M.B.; Lazzeri, A.; Martinelli, M.; Raspolli Galletti, A.M. Characterization of the *Arundo Donax* L. solid residue from hydrothermal conversion: Comparison with technical lignins and application perspectives. *Ind. Crops Prod.* **2015**, *76*, 1008–1024. [CrossRef]
56. Düdder, H.; Wütscher, A.; Stoll, R.; Muhler, M. Synthesis and characterization of lignite-like fuels obtained by hydrothermal carbonization of cellulose. *Fuel* **2016**, *171*, 54–58. [CrossRef]
57. Popescu, C.-M.; Popescu, M.-C.; Singurel, G.; Vasile, C.; Argyropoulos, D.S.; Willfor, S. Spectral characterization of *Eucalyptus* wood. *Appl. Spectrosc.* **2007**, *61*, 1168–1177. [CrossRef] [PubMed]
58. Xiros, C.; Janssen, M.; Byström, R.; Børresen, B.T.; Cannella, D.; Jørgensen, H.; Koppram, R.; Larsson, C.; Olsson, L.; Tillman, A.M.; et al. Toward a sustainable biorefinery using high-gravity technology. *Biofuels Bioprod. Bioref.* **2017**, *11*, 15–27. [CrossRef]
59. Peng, L.; Gao, X.; Chen, K. Catalytic upgrading of renewable furfuryl alcohol to alkyl levulinates using $AlCl_3$ as a facile, efficient, and reusable catalyst. *Fuel* **2015**, *160*, 123–131. [CrossRef]
60. Asghar, A.; Raman, A.; Aziz, A.; Daud, W.M.A.W. A comparison of central composite design and Taguchi method for optimizing Fenton process. *Sci. World J.* **2014**, *2014*, 1–14. [CrossRef] [PubMed]
61. Gao, X.; Peng, L.; Li, H.; Chen, K. Formation of humin and alkyl levulinate in the acid-catalyzed conversion of biomass-derived furfuryl alcohol. *Bioresources* **2015**, *10*, 6548–6564. [CrossRef]
62. Aron, M.; Rust, H. Separating off Butanol and Dibutyl Ether with the Aid of a Two-Pressure Distillation. Canadian Patent CA2227280A1; Published on 06-03-1997,
63. Naik, S.N.; Goud, V.V.; Rout, P.K.; Dalai, A.K. Production of first and second generation biofuels: A comprehensive review. *Renew. Sustain. Energy Rev.* **2010**, *14*, 578–597. [CrossRef]
64. Rakopoulos, D.C.; Rakopoulos, C.D.; Giakoumis, E.G.; Dimaratos, A.M.; Kyritsis, D.C. Effects of butanol-diesel fuel blends on the performance and emissions of a high-speed DI diesel engine. *Energy Convers. Manag.* **2010**, *51*, 1989–1997. [CrossRef]
65. Qureshi, N.; Saha, B.C.; Dien, B.; Hector, R.E.; Cotta, M.A. Production of butanol (a biofuel) from agricultural residues: Part I—Use of barley straw hydrolysate. *Biomass Bioenerg.* **2010**, *34*, 559–565. [CrossRef]

66. Koivisto, E.; Ladommatos, N.; Gold, M. Compression ignition and exhaust gas emissions of fuel molecules which can be produced from lignocellulosic biomass: Levulinates, valeric esters, and ketones. *Energy Fuels* **2015**, *29*, 5875–5884. [CrossRef]
67. Koivisto, E.; Ladommatos, N.; Gold, M. The influence of various oxygenated functional groups in carbonyl and ether compounds on compression ignition and exhaust gas emissions. *Fuel* **2015**, *159*, 697–711. [CrossRef]
68. Westbrook, C.K.; Pitz, W.J.; Curran, H.J. Chemical kinetic modeling study of the effects of oxygenated hydrocarbons on soot emissions from diesel engines. *J. Phys. Chem. A.* **2006**, *110*, 6912–6922. [CrossRef] [PubMed]
69. Ganesh, D.; Ayyappan, P.R.; Murugan, R. Experimental investigation of iso-butanol/diesel reactivity controlled compression ignition combustion in a non-road diesel engine. *Appl. Energy* **2019**, *242*, 1307–1319. [CrossRef]
70. Kumar, S.; Cho, J.H.; Park, J.; Moon, I. Advances in diesel–alcohol blends and their effects on the performance and emissions of diesel engines. *Renew. Sustain. Energy Rev.* **2013**, *22*, 46–72. [CrossRef]
71. Tree, D.R.; Svensson, K.I. Soot processes in compression ignition engines. *Prog. Energy Combust. Sci.* **2007**, *33*, 272–309. [CrossRef]
72. Rao, D.C.K.; Karmakar, S.; Basu, S. Atomization characteristics and instabilities in the combustion of multi-component fuel droplets with high volatility differential. *Sci. Rep.* **2017**, *7*, 8925. [CrossRef]
73. Lasheras, J.C.; Fernandez-Pello, A.C.; Dryer, F.L. Experimental observations on the disruptive combustion of free droplets of multicomponent fuels. *Combust. Sci. Technol.* **1980**, *22*, 195–209. [CrossRef]
74. Sluiter, A.; Hames, B.; Ruiz, R.; Scarlata, C.; Sluiter, J.; Templeton, D.; Crocker, D. *Determination of Structural Carbohydrates and Lignin in Biomass*; NREL/TP-510-42618; National Renewable Energy Laboratory: Golden, CO, USA, 2008.
75. Sluiter, A.; Ruiz, R.; Scarlata, C.; Sluiter, J.; Templeton, D. *Determination of Extractives in Biomass*; NREL/TP-510-42619; National Renewable Energy Laboratory: Golden, CO, USA, 2008.

© 2020 by the authors. Licensee MDPI, Basel, Switzerland. This article is an open access article distributed under the terms and conditions of the Creative Commons Attribution (CC BY) license (http://creativecommons.org/licenses/by/4.0/).

Article

Sustainable Exploitation of Residual *Cynara cardunculus* L. to Levulinic Acid and *n*-Butyl Levulinate

Anna Maria Raspolli Galletti [1,*], Domenico Licursi [1], Serena Ciorba [1], Nicola Di Fidio [1], Valentina Coccia [2], Franco Cotana [2] and Claudia Antonetti [1,3]

1. Department of Chemistry and Industrial Chemistry, University of Pisa, Via Giuseppe Moruzzi 13, 56124 Pisa, Italy; domenico.licursi@unipi.it (D.L.); ciorbaserena@gmail.com (S.C.); n.difidio@studenti.unipi.it (N.D.F.); claudia.antonetti@unipi.it (C.A.)
2. CIRIAF, CRB Section (Biomass Research Center), Department of Engineering, University of Perugia, Via G. Duranti 67, 06125 Perugia, Italy; valentina.coccia@unipg.it (V.C.); franco.cotana@unipg.it (F.C.)
3. CIRCC, Via Celso Ulpiani 27, 70126 Bari, Italy
* Correspondence: anna.maria.raspolli.galletti@unipi.it; Tel.: +39-50-2219290

Abstract: Hydrolysis and butanolysis of lignocellulosic biomass are efficient routes to produce two valuable bio-based platform chemicals, levulinic acid and *n*-butyl levulinate, which find increasing applications in the field of biofuels and for the synthesis of intermediates for chemical and pharmaceutical industries, food additives, surfactants, solvents and polymers. In this research, the acid-catalyzed hydrolysis of the waste residue of *Cynara cardunculus* L. (cardoon), remaining after seed removal for oil exploitation, was investigated. The cardoon residue was employed as-received and after a steam-explosion treatment which causes an enrichment in cellulose. The effects of the main reaction parameters, such as catalyst type and loading, reaction time, temperature and heating methodology, on the hydrolysis process were assessed. Levulinic acid molar yields up to about 50 mol% with levulinic acid concentrations of 62.1 g/L were reached. Moreover, the one-pot butanolysis of the steam-exploded cardoon with the bio-alcohol *n*-butanol was investigated, demonstrating the direct production of *n*-butyl levulinate with good yield, up to 42.5 mol%. These results demonstrate that such residual biomass represent a promising feedstock for the sustainable production of levulinic acid and *n*-butyl levulinate, opening the way to the complete exploitation of this crop.

Keywords: cardoon; waste biomass; hydrolysis; levulinic acid; alcoholysis; *n*-butyl levulinate; biofuels; microwaves

1. Introduction

Renewable resources have garnered increasing interest due to the shortage of petroleum, recurring rise in its price and environmental deterioration associated with its consumption, including pollutant by-products and greenhouse gases emission. The serious need to explore alternative resources respect to traditional ones for the production of chemicals and fuels has encouraged a new, more aware international policy of energy and economy. The European Union has recently incremented the renewable component of fuels and Italian legislation requires that 10% of fuels on sale in Italy are made of biofuel by 2020 [1,2]. The use of the waste biomass for energy purposes, instead of fossil fuels, should reduce the greenhouse effect, since biomass releases the same amount of carbon dioxide which has been previously trapped from atmosphere during photosynthesis. Additionally, it is worthy of notice that methane, a greenhouse gas 25 times more powerful than carbon dioxide, is emitted during the decomposition of organic material in landfills [3,4]. In this scenario, not only the simple thermal-valorization but also the chemical conversion of nonedible biomass to exploitable biomolecules should be developed and implemented, possibly avoiding any conflict with the food chain. In this regard, the ability to employ residual or waste biomasses as starting materials using water or a bioalcohol as reactants/reaction media represents an added value that fits well with the concept of green chemistry. In this context a growing interest has recently been directed to third-generation infesting plant species, such as the *Cynara cardunculus* (cardoon). It represents a promising resource to produce biomaterials and biochemicals and is a very common variety in the center of Italy and in the Mediterranean region [5]. *Cynara cardunculus* L. offers a wide spectrum of potential applications, being a rich source of fibers, oils and bioactive compounds [6,7]. Interestingly, the cultivation of such perennial herbaceous biomass shows significant advantages, such as good adaptability to climate change and growth on marginal or uncultivated lands with modest inputs, including little irrigation, care and minimal need of nutrients [8]. The seeds of the flower are exploited for oil production in food and bio-diesel supply chains. On the other hand, the nonedible lignocellulosic residues of this crop are reduced to a size of 20–40 mm by chipping and/or can undergo pretreatments that favor further exploitation of this biomass [9,10]. Steam-explosion is the most common and cost-effective method for pretreatment of lignocellulosic materials [11,12]. According to this process, chipped biomass is treated with high-pressure saturated steam for a few minutes at initial temperatures of 160–260 °C, corresponding to a pressure of 0.69–4.83 MPa, then the pressure is rapidly reduced, thus leading to explosive decompression of the biomass and fiber damage. The process causes hemicellulose degradation and lignin transformation, due to high employed temperatures, and, in addition, it reduces the cellulose crystallinity, thus increasing the effectiveness of an eventual subsequent cellulose valorization.

In the present research, two types of cardoon waste residues, remaining after seeds removal, were studied: the first one is the untreated defatted biomass, whereas the second one is the waste residue recovered after a steam-explosion pretreatment.

The hydrolysis of the cellulosic fraction of lignocellulosic biomass represents an efficient way to produce valuable platform chemicals, such as levulinic acid (LA) [13]. This last has been highlighted by the United States Department of Energy in 2010 as one of the 10 most promising building blocks in chemistry [14]. Due to its carboxyl and carbonyl functionalities, LA can be converted into various products for large-volume chemical markets, as biofuels, intermediates for chemical and pharmaceutical industries, food additives, surfactants, solvents and polymers [15]. In particular, the most promising LA derivatives for biofuel production are its alkyl esters, i.e., alkyl levulinates, γ-valerolactone, 2-methyltetrahydrofuran [16–19]. The LA yield is strongly affected by the cellulose content of the adopted feedstock [6,13] which can be changed by applying pretreatments [20]. Taking into consideration that LA is produced via hydrolysis of hexose sugars (glucose, fructose, mannose and galactose), cellulose-rich biomass proves the most suitable feedstock for its production [21].

The acid-catalyzed hydrolysis, assisted by microwave (MW) irradiation, was successfully performed for the two different samples of residual *Cynara cardunculus* L. and the effect of the main reaction parameters was investigated. Taking into account the high LA concentration reached, the hydrolysis of steam-exploded cardoon was also performed in a batch autoclave, in the perspective of a larger scaling-up. Moreover, in the same perspective of industrial applications, H_2SO_4 and HCl were selected as acid catalysts for LA synthesis due to their generally recognized advantages. They are characterized by low cost, abundant availability and high efficiency. Furthermore, they are usually employed in high TRL chemical processes, especially for the production of levulinic acid, such as the Biofine process and GF Biochemicals plants [22,23].

In addition, the direct production of *n*-butyl levulinate (BL) by alcoholysis with *n*-butanol of the cellulose-rich steam-exploded biomass was also studied and discussed, being this strategy more promising than the generally adopted procedure of esterification of neat levulinic acid [24]. Indeed, the one-pot synthesis does not require intermediate operations concerning LA concentration and purification and reduces also the waste water treatment [25–27]. BL finds interesting applications not only as solvent and intermediate, but also as valuable bioblendstock for diesel fuel, being able to reduce the emissions of particulates without increasing NO$_x$ emissions or worsening engine performance compared to neat conventional diesel fuel [27].

In Scheme 1 the two described approaches of valorization of cardoon by hydrolysis and alcoholysis are shown.

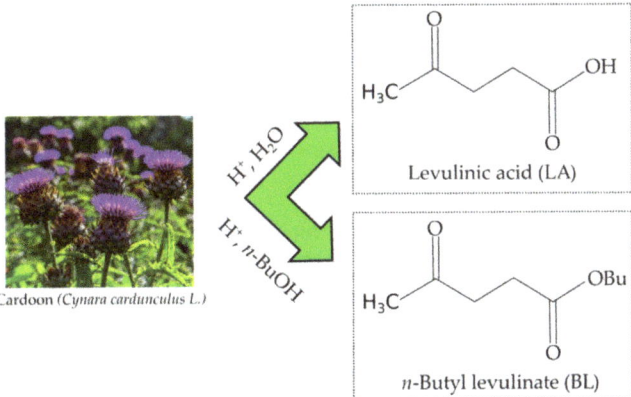

Scheme 1. Different hydrolysis and alcoholysis approaches of valorization of defatted *Cynara cardunculus* L. employed in the present research.

In the perspective of complete exploitation of the starting biomass, the solid residues recovered at the end of the hydrolysis and alcoholysis reactions were characterized by FT-IR spectroscopy, thermogravimetric and elemental analysis.

The obtained results of both hydrolysis and butanolysis runs highlight cardoon as a promising feedstock for multiproduct biorefineries. A notable example of a third-generation biorefinery fed by cardoon is the industrial site in Porto Torres (Italy). Novamont S.p.A. and ENI Versalis S.p.A. converted a petrochemical refinery into an integrated green chemicals plant, which involves local agriculture, including cultivation of cardoon and, currently, the production site is dedicated to bioplasticizers and biolubricants [28]. However, a new challenge for this innovative biorefinery is to integrate the manufacturing processes also with the synthesis of LA and ALs from the residual defatted biomasses, due to the excellent applicative perspectives of these bioproducts.

2. Results and Discussion

2.1. Compositional Analysis of Crude Cardoon Samples

The chemical compositions of the two investigated biomasses, the un-treated defatted cardoon residue (C) and the same biomass recovered after a steam-explosion pretreatment (E), are reported in Figure 1.

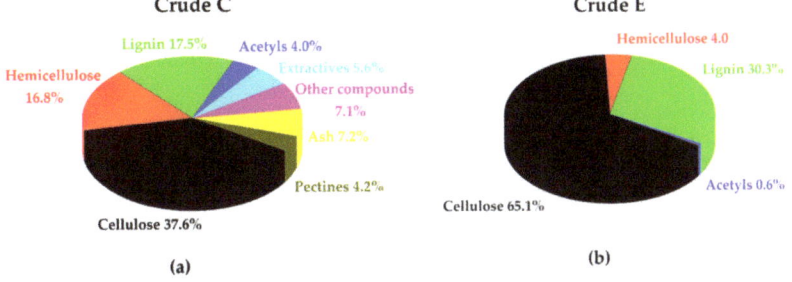

Figure 1. Chemical compositions of untreated cardoon residue (**a**) and the same biomass recovered after a steam-explosion pretreatment (**b**), reported as wt%, on dry basis.

The steam-explosion pretreatment is a high-pressure and high-temperature physical process that has been widely used for biomasses in the last decades [29], since it allows increased availability of cellulose of the raw lignocellulosic matrix for further processing. The possible use of steam-exploded pretreated biomass is on one hand the production of bioethanol or more in general energy carriers [30] and on the other hand the possibility to obtain high value biochemical, such as the nanocrystalline cellulose [5] and other products useful for several chemical or pharmaceutical applications. The input matrices to the steam-explosion pretreatment section can be lignocellulosic softwoods, hardwoods and spruces but also residues obtained from agroindustrial activities, such as tomatoes, vineyard pruning or residues of infesting species, such as the *Cynara cardunculus* that can be considered third-generation biomass of growing interest for the production of sustainable biomaterials [31]. The process "intensity" is generally described by the severity factor (R0) that considers the process parameters, such as the temperature and the pressure. In the case of lignocellulosic materials, the process yields can be significantly increased using a double-step process of both solid and liquid fraction after the pretreatment [30]. The steam-explosion facility used for this research is available at the University of Perugia (CRB/CIRIAF) and it is composed of: (i) vapor generator; (ii) charging section for raw biomass; (iii) expansion valves; (iv) high-pressure reactor; (v) postexplosion tank and (vi) exploded liquid recovery section [30].

Both C and E samples show very different content of cellulose, about 38 and 65 wt%, respectively. In fact, after the steam-explosion pretreatment, the hemicellulose amount decreased from about 17 to 4 wt%, whereas extractives and ash were removed, as expected. On the other hand, lignin increased from about 17 up to 30 wt%, due to the reduction of the content of other components. The above-reported composition was determined on the dry biomasses and the enrichment of cellulose content in the steam-exploded sample proved very important from an industrial point of view because at equal biomass loading, it enables the processing of a higher amount of cellulose, which results in marked increases of target product concentrations in the reaction mixture with subsequent ease for the successive work-up, separation and purification processes. The available defatted raw sample C had a humidity amount of 5.9 wt%, while, as expected, the steam-exploded sample E had a humidity level of 73.6 wt%. Both wet as-received samples and dried ones were tested in the catalytic runs.

2.2. MW-Assisted Hydrolysis of C and E Cardoon Samples

A preliminary study on the effect of the catalyst types, their amount and humidity on LA formation was performed. The hydrolysis of dry cardoon, both C and E samples, was carried out using two acid catalysts, HCl and H_2SO_4. The amount of these mineral acids was calculated to have the same concentration of hydronium ions in the starting mixtures. Table 1 reports the results for the experiments performed employing a biomass loading of 10 wt% (on dry basis), working at 190 °C for 20 min in the MW reactor.

Table 1. Hydrolysis experiments of C and E cardoon samples with different type and amount of catalysts and humidity grade. Experimental conditions: 190 °C, 20 min, biomass loading = 10 wt% (on dry basis), MW heating.

Run	Catalyst (wt%)	Sub/Cat (mol/mol) [a]	Products (g/L) [b]				LA Ponderal Yield (wt%)	LA Molar Yield (mol%)
			Glu	AA	FA	LA		
C1 (dry) [c]	HCl (1.6 wt%)	0.9	0.2	5.6	8.2	15.5	13.3	49.3
C2 (dry) [c]	H_2SO_4 (2.1 wt%)	1.8	0.1	6.0	7.1	13.0	11.2	41.5
E1 (dry) [c]	HCl (1.6 wt%)	0.9	0.1		7.9	26.9	23.0	48.4
E2 (dry) [c]	H_2SO_4 (2.1 wt%)	1.8	0.5		7.1	22.5	19.3	40.7
C3 (wet) [d]	HCl (1.6 wt%)	0.9	0.2	5.6	8.5	15.9	13.6	50.7
E3 (wet) [d]	HCl (1.6 wt%)	0.9	0.1		9.4	25.5	23.6	49.7

[a] Substrate to catalyst molar ratio: mol of anhydrous glucose unit in the starting biomass/mol of catalyst; [b] Glu = glucose; AA = acetic acid; FA = formic acid; LA = levulinic acid; [c] the sample was used after drying step; [d] the sample was employed as received.

In the presence of HCl, both dry samples (runs C1 and E1, Table 1) achieve comparable LA yields, 49.3 and 48.4 mol% respectively. Adopting the same dry samples, when H_2SO_4 was used (runs C2 and E2), lower LA molar yields were obtained, 41.5 and 40.7 mol% for C and E cardoon samples, respectively. As reported in previous studies, HCl enhances LA hydrolysis via a one-pot mechanism, since Cl^- ions catalyze the intermediate 5-hydroxymethyl-2-furaldehyde (HMF) rehydration reaction, while SO_4^{2-} ions are responsible for an inhibitor effect [32]. In all these runs, low amounts of unconverted glucose were detected, while no intermediates, such as furfural and HMF, were observed. On the other hand, appreciable amounts of formic acid (FA), which is another commodity chemical [8], were coformed by the hydrolysis of both C and E cardoon samples. Indeed, FA is coproduced during biomass hydrolysis [33] and its use in several fields has further encouraged the interest in LA synthesis. In our experiments, FA formation seems to be slightly facilitated when HCl is employed (compare runs C1 with C2 and runs E1 with E2) and for this reason, only HCl was selected for the subsequent investigation regarding the humidity grade. To investigate this parameter, both the cardoon samples C and E were adopted as-received with the humidity grade of 5.9 and 73.6 wt% respectively (runs C3 and E3), using the same experimental conditions of runs C1 and E1. As expected, dry and wet cardoon samples show similar LA molar yields, which are 49.3 and 50.7 mol% for dry and wet C (runs C1 and C3), respectively, and 48.4 and 49.7 mol% for dry and wet E (runs E1 and E3), respectively. Based on these results, since wet biomass can be directly used, only as-received wet cardoon samples were employed for a more detailed investigation. In the perspective of industrial application, the adoption of a high biomass loading is to be preferred, thus applying the high-gravity approach to achieve the highest products concentration. In other words, this is the same concept already applied regarding the cellulose content (on dry biomass) with the steam-explosion pretreatment. The high-gravity method presents several advantages for an industrial perspective: it increases the concentration of crude products, reduces the costs for their purification and waste-water treatment. On the other hand, in an over-loaded reactor, a more difficult physical agitation of the reaction slurry could lead to a significant decrease in the reaction rate. Table 2 reports the results of hydrolysis experiments carried out by increasing the biomass loading from 10 up to 20 wt% on dry basis, keeping constant the other reaction parameters (190 °C, 20 min, HCl as catalyst). In the same Table 2, the effect of the substrate to catalyst ratio was investigated at 20 wt% of biomass loading, for both wet C and E cardoon samples.

Table 2. Hydrolysis experiments on wet C and E cardoon samples with different biomass loadings and substrate/catalyst molar ratios. Experimental conditions: 190 °C, 20 min, HCl as catalyst, MW heating.

Run	Biomass Loading (wt%) [a]	Sub/Cat (mol/mol) [b] (HCl wt%)	Products (g/L) [c]				LA Ponderal Yield (wt%)	LA Molar Yield (mol%)
			Glu	AA	FA	LA		
C3	10	0.9 (1.6 wt%)	0.2	5.6	8.5	15.9	13.6	50.7
C4	15	0.9 (2.4 wt%)	0.5	8.6	12.7	23.9	13.0	48.1
C5	20	0.9 (3.2 wt%)		9.3	12.7	34.6	13.0	48.5
C6	20	1.5 (1.9 wt%)		9.7	13.4	27.8	10.7	39.8
C7	20	2.0 (1.4 wt%)		7.5	12.7	34.3	13.3	49.6
E3	10	0.9 (1.6 wt%)	0.1		9.4	25.5	23.6	49.7
E4	15	0.9 (2.4 wt%)			11.7	47.0	24.6	51.8
E5	20	0.9 (3.2 wt%)			12.5	58.9	21.1	44.5
E6	20	1.5 (1.9 wt%)			15.9	55.4	20.7	43.6
E7	20	2.0 (1.4 wt%)			20.4	59.0	22.4	47.3

[a] determined on dry basis; [b] substrate to catalyst molar ratio: mol of anhydrous glucose unit in the starting biomass/mol of catalyst; [c] Glu = glucose; AA = acetic acid; FA = formic acid; LA = levulinic acid.

For both C and E samples, no significant decrease of the yields was observed increasing the biomass loading. On the other hand, higher LA concentrations, from 15.9 up to 34.6 g/L for C samples, and from 25.5 to 58.9 g/L for E ones, were reached. Considering that lower concentrations of HCl allow minimizing environmental impact and process costs, Table 2 reports the results of experiments performed with lower HCl amounts, increasing the substrate/catalyst ratios from 0.9 up to 2.0 mol/mol, run C5, C6, C7 and runs E5, E6 and E7, for C and E samples respectively. These runs evidenced that LA molar yield is not affected by the reduction of the catalyst content, at least in the range of the investigated substrate to catalyst ratios. For both C and E samples, promising values of LA molar yields were obtained, as well as no significant formation of side-products. In the case of E biomass hydrolysis, even acetic acid was not observed, due to the low content of acetyl groups in this feedstock. As expected, higher LA concentrations were achieved from the cellulose-rich steam-exploded cardoon (E sample) rather than from that untreated (C sample), justifying the adopted pretreatment in the industrial perspective: the maximum reached value was 59.0 g/L for E and 34.6 g/L for C. On the other hand, it is interesting that the reactivity of the cellulosic fraction of the exploded biomass E is very similar to that of the untreated sample C, thus suggesting that pretreatment does not significantly modify the accessibility of the cellulose fibers. Regarding FA, its formation is almost equimolar with respect to LA for runs C3–E3 of Table 2, as expected from the overall hydrolysis reaction mechanism, whilst this correspondence does not fully fit for the remaining runs E4–E7 of Table 2 (lower FA molar concentration than the theoretical one), probably due to a combined effect of the type of feedstock, which has been previously steam-exploded (resulting more reactive than the crude sample to the hydrolysis), and of the high loading used for these runs, both leading to its more appreciable thermal degradation to CO_2 and H_2 in the liquid phase.

At the end of every run, a solid residue was recovered, accounting for about 30 wt% respect to the starting dry biomass. Due to the ponderal relevance of these residues, their characterization and applicative perspectives will be discussed later.

Regarding the presence of by-products, the reaction mixture certainly includes also low amounts of soluble impurities, which have not been considered up to now and which should be otherwise better characterized, to get information about the proper work-up procedures, although the concentration of any single by-product in the hydrolyzates is under the limit of detection of the HPLC analysis by routine refractive index detector (0.1 g/L) and, for this reason, not inserted in Tables 1 and 2. To detect trace amounts of impurities, higher sensitivity mass and/or UV detectors should be used. For this purpose, first of all, the crude hydrolysate obtained from run E7 of Table 2 was extracted by diethyl ether and the recovered extract was analyzed by the GC-MS technique. The analysis revealed the presence of oxygenated C5 compounds and of aromatics of lignin sources, such as guaiacol (2-methoxy phenol) and syringol (2,6-dimethoxy phenol) (see Figure S1, Supplementary Section,). The dehydration of C5 and C6 carbohydrates causes the formation of furanic monomers and soluble precursors of solid humins [34]. On the other hand, the above phenolic derivatives show characteristic absorptions in the ultraviolet region, in particular at 284 nm, due to carbonyl n π* transitions [35], while stronger carbonyl π π* transitions occur at a lower wavelength, at about 205 nm. Based on these statements, the HPLC-UV analysis of the crude hydrolysate was carried out at 284 and 205 nm, thus better differentiating furanic impurities from other carbonyl species, such as aliphatic carboxylic acids. Regarding the UV-HPLC chromatogram at 284 nm (Figure S2, Supplementary Section), it shows the presence of many furanic/aromatic species, which elute after 20 min, whereas at 205 nm aliphatic carbonyl species prevail (Figure S3, Supplementary Section). To obtain more in-depth information, the crude liquor was analyzed by HPLC-MS and total ion current (TIC) chromatogram related to all ions of all detected masses and UV chromatograms at 280, 250 nm and 205 nm are reported in the Supplementary Section, together with the chromatographic data of the main detected compounds including, where possible, the best matched chemical formula (Figure S4 and Table S1, respectively). At this level of investigation, given the high sensibility of this technique, the TIC signal due to the crude hydrolysate is very complex, including many impurities, besides LA (compound n° 17 of Table S1, Supplementary Section), which by far represents the main product of interest, as previously stated,. The cross-comparison with the GC-MS data confirms the presence of methoxyphenol and 2,6-dimethoxyphenol (compounds n° 36 and n° 38 of Table

S1 of Supplementary Section, respectively), as aromatics of lignin source. To obtain further useful information for the following discussion, some typical compounds deriving from biomass hydrolysis have been sought [36], paying particular attention to some selected carboxylic acids of interest [37]. On this basis, after having excluded the contribution of the LA as the main component, extracted ion chromatogram (XIC) of some typical organic acids deriving from biomass hydrolysis treatment were acquired (Figure S5/Table S2 and Figure S6/Table S3 of Supplementary Section), thus ascertaining the presence, in low abundance, of many of these, such as tartaric, malic, succinic, lactic, butyric, itaconic, maleic, pyruvic, glutaric, adipic, 2-hydroxy-2-methylbutyric, gluconic and citric acid. In addition, 5-HMF is still present, as well as its dimeric/trimeric derivatives, as confirmed by the corresponding XIC processing (Figure S7, Supplementary Section). Instead, furfural, the furanic compound deriving from the acid-catalyzed conversion of the hemicellulose fraction, has not been found, thus indirectly confirming the effectiveness of the steam-explosion treatment, which has allowed the preliminary removal of this biomass component. Lastly, glucose has been identified only in traces (Figure S5/Table S2 of Supplementary Section), confirming the effectiveness of the acid-catalyzed conversion of the C6 fraction, as highly desired by our approach. Lastly, in order to demonstrate the cost-effectiveness of the hydrolysis process, the purification of the crude hydrolysate deriving from run E7 of Table 2 was carried out on laboratory scale adopting 2-methyltetrahydrofuran as extraction solvent, followed by subsequent fractional distillation of the corresponding extract, according to the experimental procedure reported in the Materials and Methods section. By this way, the final isolated LA yield of 20.3 wt% was ascertained, similar to that obtained from the HPLC analysis of the corresponding crude hydrolysate (run E7 of Table 2). The ascertained LA purity grade was 93%, estimated by both the GC and HPLC techniques, including FA, AA and angelica lactone as residual impurities. The latter compound originates from the acid-catalyzed LA dehydration, followed by ring closure, which typically occurs upon LA heating, during the purification procedure.

Steam-exploded cardoon E was selected for further studies regarding the effect of temperature on the hydrolysis reaction and Table 3 reports the results achieved at the lower temperature of 180 °C and the upper one of 200 °C, in comparison with the run at the temperature of 190 °C previously used. The obtained results confirm that 190 °C is the best temperature for the production of LA, as highlighted in the literature for the hydrothermal treatment of many lignocellulosic biomasses [13].

Table 3. Hydrolysis experiments on wet E cardoon sample, with different temperatures. Experimental conditions: reaction time = 20 min, sub/cat = 2.0 mol/mol, biomass loading = 20 wt% (on dry basis), HCl as catalyst (1.4 wt%), MW heating.

Run	Temperature (°C)	Products (g/L) [a]				LA Ponderal Yield (wt%)	LA Molar Yield (mol%)
		Glu	AA	FA	LA		
E8	180			19.4	46.4	18.6	39.8
E7	190			20.4	59.0	22.4	47.3
E9	200			13.8	48.0	19.2	41.2

[a] Glu = glucose; AA = acetic acid; FA = formic acid; LA = levulinic acid.

Finally, the effect of the reaction time on LA formation was investigated (runs E7, E10 and E11, Figure 2). As reported in Figure 2, experiments were carried out on E sample at 20, 40 and 60 min of MW irradiation, under the same reaction conditions of temperature (190 °C), catalyst (HCl, 1.4 wt%), substrate-to-catalyst molar ratio (2.0 mol/mol) and biomass loading (20 wt%).

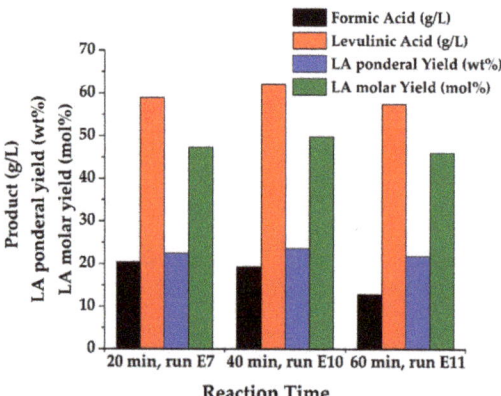

Figure 2. Hydrolysis experiments on wet E cardoon sample at different reaction times. Experimental conditions: 190 °C, HCl as catalyst (1.4 wt%), sub/cat = 2.0 mol/mol, biomass loading = 20 wt% on dry basis, MW heating.

Only a weak time effect on the hydrolysis efficiency was observed, achieving the best LA yield of 49.9 mol% with the highest LA concentration of 62.1 g/L after 40 min of heating, adopting HCl in a low amount, 1.4 wt%.

2.3. Hydrolysis of E Cardoon under Conventional Heating

Based on the results discussed in the previous section, cellulose-rich steam-exploded cardoon (E sample) proved to be a particularly promising residual biomass for LA synthesis. Therefore, this hydrolysis reaction was also studied in a batch autoclave with conventional heating to verify the possible process intensification on an industrial scale. Taking into account that the heat transfer process in the autoclave is slower than the heating mechanism in the MW reactor, a longer reaction time (120 min) was applied in the autoclave, adopting two different biomass loadings, 10 and 15 wt% on dry basis. For comparison, MW hydrolysis reactions had been carried out for 20 and 40 min, adopting the same biomass loadings. The results and the compositions of the main reaction products are reported in Table 4 and Figure 3, respectively.

Table 4. Hydrolysis experiments on wet E cardoon sample adopting different heating systems, biomass loadings and substrate to catalyst molar ratios. Experimental conditions: 190 °C, HCl as catalyst.

Run	Biomass Loading (wt%) [a]	Sub/Cat (mol/mol) [b] (HCl wt%)	Heating System	Time (min)	LA Ponderal Yield (wt%)	LA Molar Yield (mol%)
E3	10	0.9 (1.6 wt%)	MW	20	23.6	49.7
E12	10	0.9 (1.6 wt%)	autoclave	120	25.2	53.1
E13	15	1.5 (1.4 wt%)	MW	40	24.6	51.8
E14	15	1.5 (1.4 wt%)	autoclave	120	22.3	47.1

[a] Determined on dry basis; [b] substrate to catalyst molar ratio: mol of anhydrous glucose unit in the starting biomass/mol of catalyst.

In the case of biomass loading of 10 wt%, similar LA molar yields were reached with the two heating systems, 49.7 and 53.1 mol% in MW and autoclave, respectively, while in the case of 15 wt% of biomass loading, heating in autoclave resulted in slightly lower yield, 47.1 mol%, compared to 51.8 mol% in MW, probably due to mass and heat transfer limitations. However, the comparable high LA concentrations reached both in the autoclave and in MW, 41.4 and 45.5 g/L respectively, demonstrate that both heating methodologies can be successfully applied for performing this process.

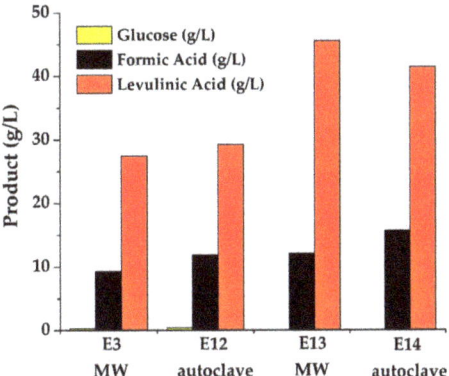

Figure 3. Compositions of the main reaction products of runs E3, E12, E13 and E14 reported in Table 3.

2.4. MW-Assisted Alcoholysis of E Cardoon Sample

In the second part of our investigation, the one-pot alcoholysis reaction of wet E cardoon in *n*-butanol to BL was studied in the MW reactor and the results are reported in Table 5. The presence of a significant amount of introduced water, due to the high humidity of the sample, is an unprecedented approach respect to the up to now reported alcoholysis studies, performed in the presence of alcohol alone as reactant/reaction medium. This procedure allowed us to convert efficiently the starting biomass, without the initial drying step, thus at the same time-saving time and significant resources.

Table 5. Butanolysis experiments on wet E cardoon sample, adopting different biomass loadings and catalyst amounts. Experimental conditions: 190 °C, H_2SO_4 as catalyst, MW heating.

Run	Biomass Loading (wt%) [a]	Sub/Cat (mol/mol) [b] (H_2SO_4 wt%)	Time (min)	BL (g/L)	BL Ponderal Yield (wt%)	BL Molar Yield (mol%)
AE1	8	2.4 (1.3 wt%)	15	27.5	26.4	42.5
AE2	15	2.4 (2.4 wt%)	15	44.2	15.3	22.1
AE3	15	4.7 (1.3 wt%)	15	37.9	12.8	19.8
AE4	8	2.4 (1.3 wt%)	30	22.0	25.3	36.6
AE5	8	2.4 (1.3 wt%)	45	20.9	24.2	35.1

[a] Determined on dry basis; [b] substrate to catalyst molar ratio: mol of anhydrous glucose unit in the starting biomass/mol of catalyst.

In the presence of low contents of the acid catalyst H_2SO_4 (1.3 and 2.4 wt%) with the same substrate to catalyst molar ratio, and applying only 15 min of heating, the butanolysis led to BL molar yields of 42.5 and 22.1 mol%, using respectively the biomass loading of 8 and 15 wt% on dry basis, runs AE1 and AE2 (Table 5). When the lower catalyst loading of 1.3 wt% was maintained with the biomass loading of 15 wt%, a significant decrease of BL yield (19.8 mol%) was ascertained, essentially due to the above-mentioned mass transfer issues. Taking into account that the best BL yield, as well as a good BL concentration (27.5 g/L), were obtained in the run AE1, the effect of duration was studied adopting the same conditions but prolonging the reaction time to 30 and 45 min (runs AE4 and AE5, respectively). The achieved results show that the prolonging of the reaction time does not improve the BL production, whereas slightly lower BL yields and BL concentrations were obtained, demonstrating the efficacy of MW irradiation within a short reaction time. Finally, to better evaluate the impact of introduced water on the ascertained performances, an explorative run adopting the dry cardoon E sample was performed employing the same reaction conditions of run AE1: in this case, the BL concentration of 17.9 g/L was achieved with BL ponderal and molar yields of 26.4 wt% and 38.3 mol% respectively, proving similar to that ascertained on wet cardoon. In all the above runs only traces of levulinic acid were detected, thus confirming that under an excess of bioalcohol the hydrolysis reaction is irrelevant and also when a certain humidity is introduced with the wet biomass.

At the end of the reaction for every run, a significant amount of solid residue was also ascertained, whose characterization and potential valorization will be discussed below.

The results of this preliminary study demonstrate that direct BL production can be performed not only from conventional starting materials (LA, disaccharides, polysaccharides and furfuryl alcohol [24]) but also directly, using raw steam-exploded defatted cardoon as the starting feedstock, opening the way to the direct conversion of the cellulosic fraction of this cheap, residual biomass in a valuable intermediate/bio-fuel.

2.5. Characterization of Postreaction Solid Residues

The solid residues recovered from the best MW-assisted hydrolysis and alcoholysis reactions to LA and BL, runs E7 and C7 for hydrolysis and run AE1 for alcoholysis, amounted respectively 30.6, 31.4 and 27.4 wt% of the starting biomass, calculated on a dry basis. All these samples were analyzed by elemental analysis and Table 6 reports the obtained results, compared with the C and E dry cardoon samples starting feedstocks.

Table 6. Results of the elemental analysis for the starting C and E dry cardoon samples, for the solid residues at the end of runs C7 and E7 (hydrolysis reactions) and for the solid residue at the end of run AE1 (alcoholysis reaction).

Sample	C (%)	H (%)	N (%)	S (%)	O (%) [a]	Ash (%)	H/C	O/C	HHV (MJ/kg)
Cardoon C—starting feedstock	43.5	6.4	0.2	0.3	49.6	7.2	1.8	0.9	17.47
Cardoon E—starting feedstock	47.8	6.4	0.2	0.2	45.4	0.0	1.6	0.7	19.57
Solid residue, run C7	66.9	5.1	0.1	0.1	27.8	0.2	0.9	0.3	26.49
Solid residue, run E7	66.4	5.2	0.1	0.1	28.2	0.3	0.9	0.3	26.39
Solid residue, run AE1	65.3	5.2	0.2	0.1	29.2	0.3	0.9	0.3	25.90

[a] Oxygen content was calculated by difference: O (%) = 100 (%)–C (%)–H (%)–N(%)–S(%).

The above data related to the starting feedstocks C and E confirm only a limited beginning of carbonization for the latter sample, as shown by the slight increase in its carbon content, occurred as a consequence of the mild steam-explosion treatment, aimed at the breakdown of the biomass matrix (cross-linking lignin), the bulk solubilization of the hemicellulose fraction and the removal of smaller hydrocarbon molecules (volatiles and gases) [38]. Instead, more advanced carbonization has occurred as a consequence of the acid-catalyzed hydrothermal treatment. H/C and O/C molar ratios of both solid residues at the end of runs C7 and E7 fall within the range reported in the literature for the hydrochars (H/C: ~0.8–1.4 and O/C: ~0.3–0.5) [39] and also in agreement with our previous work [40]. Therefore, the differences in the carbon content of the two different starting feedstocks (cardoon C and E, both as starting feedstocks in Table 6) have been attenuated by the acid-catalyzed hydrothermal treatment (solid residue of run C7 and E7, respectively, in Table 6), demonstrating the progressed carbonization occurred with this technology. Lastly, the postalcoholysis residue (solid residue of run AE1 in Table 6) does not show significant compositional differences respect to the posthydrolysis ones (solid residues of run C7 and E7 in Table 6), thus highlighting the similarity between the performed hydrolysis/alcoholysis treatments. These conclusions are also evident from the Van Krevelen diagram shown in Figure 4 which plots the molar ratios of the H/C and O/C for both the starting biomasses, cardoon samples C and E, and for the solid residues at the end of runs C7 and E7 for hydrolysis and run AE1 for butanolysis.

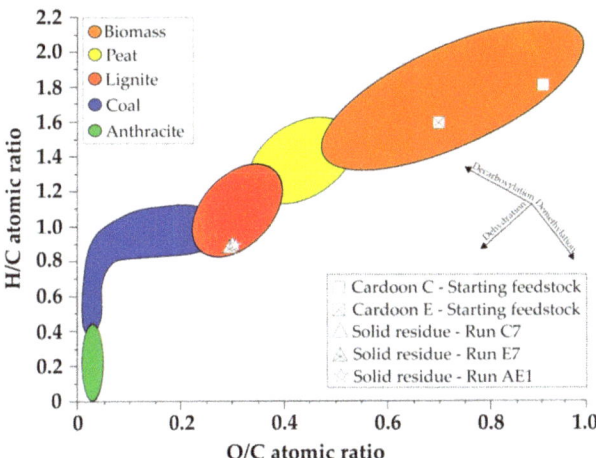

Figure 4. Van Krevelen diagram of starting biomasses, cardoon sample C7 and E7, and solid residues at the end of runs C7 and E7 for alcoholysis and run AE1 for butanolysis.

The positions of the starting feedstocks and the corresponding chars in the Van Krevelen diagram confirm that, in all cases, dehydration is the main allowed path, leading to the formation of carbonaceous material and the H/C and O/C ratios of all the obtained residues fall within the range reported in the literature for this kind of biomaterial (H/C: ~0.8–1.4 and O/C: ~0.3–0.5) [40,41]. In addition, the higher heating value (HHV) was calculated from elemental analysis (Table 6): the ascertained values significantly increase going from the starting feedstock to the corresponding chars, thus resulting comparable with that of the traditional lignite coal, in agreement with the conclusions gathered from the Van Krevelen plot [41].

Furthermore, FT-IR characterization of the postreaction solid residues was performed and Figure 5 shows the FT-IR spectra registered in ATR mode both for the starting crude cardoon, both C and E samples, and for the solid residues recovered after hydrolysis reactions, runs E7 and C7.

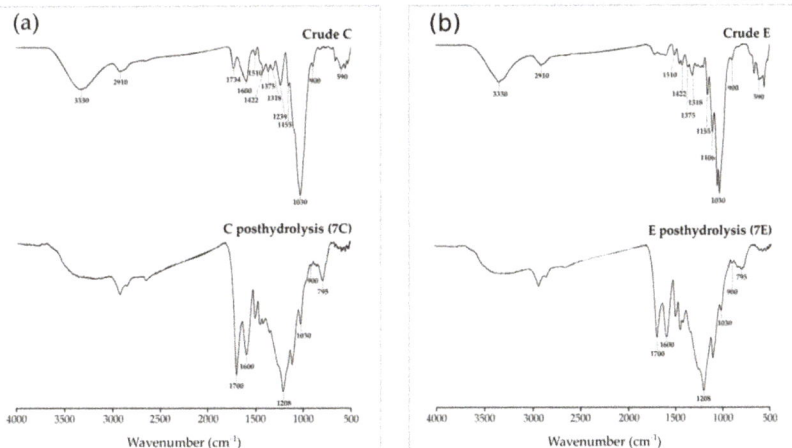

Figure 5. FT-IR spectra registered in ATR mode for the starting crude cardoon, both C and E samples, and for the solid residues recovered after hydrolysis reactions, runs E7 and C7.

The spectrum of crude sample C shown in Figure 5a shows characteristic signals of the three biopolymers, cellulose, hemicellulose and lignin: a broad band at about 3330 cm^{-1} assigned to vibration mode of O–H bonds and a band at 2910 cm^{-1} assigned to C–H stretching of methyl and methylene groups. The peak at 1734 cm^{-1} is due to C=O stretching in the acetyl group and carboxylic acid of hemicellulose, while the signals at 1600 and 1510 cm^{-1} are attributed to aromatic C=O stretching and C=C vibration of lignin [41,42]. Weak peaks are observed at 1422, 1375 and 1318 cm^{-1} that can be assigned to the C–H asymmetric mode of CH$_2$ in cellulose, aromatic C–H and C–O in lignin, respectively [43,44]. The band at 1239 cm^{-1} is due to the C–O stretching of alcoholic, phenolic and ether groups [44]. The peak at 1155 cm^{-1} is assigned to C–O–C asymmetric stretching in cellulose and hemicellulose, the intense peak at 1030 cm^{-1} to C–O–H stretching and the one at 900 cm^{-1} to anomeric vibration at the β-glycosidic linkage, while the band at 590 cm^{-1} can be due to aromatic C–H bonds [40,45]. As expected, the peaks at 1734 and 1600 cm^{-1} are not noticeable in the spectrum of crude sample E showed in Figure 5b, since the steam-explosion pretreatment degrades hemicellulose and lignin. The FT-IR spectroscopy in ATR mode allows the characterization of functional groups on the material surface and the analysis is in agreement with the bulk composition determined by the NREL procedure for the samples of raw cardoon. In the case of the solid residues recovered after the posthydrolysis reactions 7E and 7C, reported again in Figure 5a,b respectively, peaks characteristic of cellulose, such as the intense peak at about 1030 cm^{-1} and the peak at about 900 cm^{-1}, are strongly decreased in accordance with the fact that cellulose was efficiently converted to LA. However, signals that may be assigned to byproducts such as humins appear: a peak at about 1700 cm^{-1} due to the C=O stretching, another one at about 1600 cm^{-1} due to stretching vibration of C=C bonds of furanic rings and the last one at about 795 cm^{-1} due to aromatic bending off the plane of the C–H bond [38,44] are evident in the spectrum of both post-hydrolysis residues. Broad absorbance in the 1300–1100 cm^{-1} region could be ascribed to multiple C–OH stretching bonds. However, the intense signal at 1208 cm^{-1} evident in the posthydrolysis residues 7C and 7E could arise also from the presence of ether bonds [46]. Humins can result from condensation reactions between sugars, HMF and intermediates during the dehydration of carbohydrates [47–50] and their formation can be competitive with the rehydration of HMF to LA. Moreover, humins are complex and recalcitrant carbonaceous materials that can cover the substrate surface, making it less accessible to acid attack.

Figure 6 reports the FT-IR spectra of the solid residue recovered after the butanolysis reaction AE1 together with the starting crude E sample.

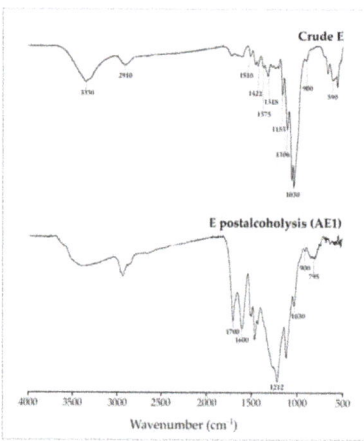

Figure 6. FT-IR spectra registered in ATR mode for the starting cardoon E sample and that of the solid residue recovered after alcoholysis reaction AE1.

In Figure 6, the FT-IR spectrum of the postalcoholysis char is shown, resulting very similar to that obtained after hydrothermal processing (Figure 5). In particular, the absorption bands at 1700 and 1600 cm^{-1} (C=O and C=C stretching vibrations, respectively) are visible also in the postalcoholysis char, as well as that at 1030 cm^{-1} (C–O–H stretching in cellulose), even if in this case of lower intensity respect to the crude E feedstock, thus confirming the occurred cellulose decomposition, as previously stated. In addition, the absorption bands at 1208 cm^{-1} (C–O stretching), and 795 cm^{-1} (C–H bending off the plane) further confirm the similarity between the chars produced by hydrolysis/alcoholysis. At this stage of the investigation, the severity of the performed treatments dampens any differences deriving from the use of different reaction solvents, otherwise reported by some authors very recently [51].

Moreover, the recovered residues and the starting biomasses were also characterized by thermogravimetric analysis and weight loss and weight loss thermograms are depicted in the Supplementary Section (Figure S8).

Regarding the devolatilization behavior of the starting feedstocks (Figure S8, thermograms (a) and (b)), the first peak is found below 100 °C and is due to the loss of humidity. The degradation of the cardoon starts at higher temperatures. In particular, a shoulder is found at about 250 °C, present only in the C cardoon starting feedstock, ascribed to hemicellulose fraction, overlapped with that of cellulose, in the range 300–350 °C [52,53]. Lastly, lignin degradation is very slow, occurring for the whole temperature range of the thermogravimetric analysis [52]. The comparison between the thermograms of the starting feedstocks, C and E cardoon samples ((a) and (b)), confirms the effectiveness of the steam-explosion treatment, which has allowed the selective removal of the hemicellulose fraction. Moreover, the E cardoon starting feedstock shows a very weak shoulder at about 400 °C, which is due to the lignin component [52], which proved more accessible and reactive as a consequence of the occurred steam-explosion pretreatment. On the other hand, the three residues deriving from hydrolysis/alcoholysis treatments ((c), (d), (e)) show analogous thermal profiles, thus confirming their chemical similarity, also in agreement with the previous characterization data. Again, the humidity loss of these samples occurs below 100 °C, then a degradation step was found at about 200 °C, due to the release of some organic compounds (such as LA and FA), trapped into this porous bio-material [54]. The absence of the degradation steps of hemicellulose and cellulose fractions confirms the effectiveness of the performed hydrolysis/alcoholysis treatments, whereas the main peak at about 400 °C is attributed to the decomposition of volatile lignin/furanic structures [54,55]. In all cases of the synthesized residues, a final weight loss of ~50 wt% was ascertained, thus revealing increased thermal stability above the starting biomass and demonstrating similar carbonization occurred [40], also in agreement with the previous characterization data.

The characterization of the obtained solid residues opens the way to their uses. Regarding this aspect, certainly the most immediate use is the combustion for the energy recovery, but this choice is currently considered as the last option, preferring, when possible, its reuse within the scopes of the circular economy [56]. For agricultural uses, the application of biochar to the soil can mitigate climate change by promoting carbon sequestration and decrease greenhouse gas emissions [57]. Moreover, char has been advantageously proposed as a growing medium, to be used in combination with other components (vermicultite, clays, etc.) to improve physicochemical soil properties, such as increase in cation exchange capacity, water holding capacity, available water, improvement of soil structure, reduction in soil acidity, microbiological activity, quality, and yield of the crops [57]. Use of the char as a soil amendment for the recovery of contaminated soils, including stabilization of organic and inorganic contaminants, has been proposed and seems attractive [57]. Lately, more added-value char-based products are under development in many research fields, such as adsorption, catalysis and electrochemical energy storage (lithium-ion batteries, lithium-sulfur batteries, sodium-ion batteries and supercapacitors), after having properly tuned its porosity and functional groups, by choosing the appropriate starting feedstocks and optimizing the reaction conditions [58].

3. Materials and Methods

3.1. Materials

Two types of cardoon waste residues after seeds removal were investigated: non-pretreated cardoon (C) and steam-explosion pretreated cardoon (E). Catalysts and chemicals were purchased from Sigma-Aldrich and employed as received: hydrochloric acid (HCl, 37 wt%), sulfuric acid (H_2SO_4, 95 wt%), 5-hydroxymethyl-2-furaldehyde (HMF, 98%), levulinic acid (LA, 98%), formic acid (FA, 98%), glucose (Glu, 99.5%), acetic acid (AA., 99%), furfural (99%), diethyl ether (98%), water for HPLC; n-butanol (94,5%), n-dodecane (99%), n-butyl levulinate (BL, 98%). Both C and E cardoon samples were used as received and/or after a drying step carried out at 105 °C in an oven until a constant weight was reached.

The biomass loading and the catalyst amount were calculated according to the following equations, respectively:

Biomass loading (wt%) = employed dry biomass (g)/[employed biomass (for dry or wet samples) (g) + solvent (g)] × 100;

Catalyst (wt%) = catalyst (g)/[catalyst (g) + employed biomass (for dry or wet samples) (g) + solvent (g)] × 100.

3.2. Steam-Explosion Process Conditions

The steam-explosion pretreatment was carried out on the untreated Cynara cardunculus L. sample C using the CRB/CIRIAF equipment [30]. In particular, 447.50 g of dry cardoon sample was treated at 165 °C and 200 bar for 10 min, employing the severity factor Log R0 equal to 2.91.

3.3. Compositional Analysis of Raw Biomass

For both C and E cardoon residues, the contents of cellulose, hemicellulose and lignin were determined according to the NREL protocol, as well as extractives and ash [59–61]. The content of humidity was estimated according to the NREL procedure [62].

3.4. Acid-Catalyzed Hydrolysis of Cardoon

Hydrolysis reactions were carried out in deionized water, using HCl or H_2SO_4 as catalyst, in the single-mode MW reactor CEM Discover S-class System (maximum pulsed-power 300 W, 35 mL pyrex vial). The reaction slurries were mixed with a magnetic stirrer and irradiated up to the set-point temperature for the selected rection time. During the reaction, pressure and temperature values were continuously acquired with the software and controlled with a feedback algorithm to maintain the constant temperature. Batch experiments were carried out also in an electrically heated 600 mL Parr zirconium made-fixed head autoclave equipped with a P.I.D. controller (4848). The reactor was pressurized with nitrogen up to 30 bar and the reaction mixtures were stirred using a mechanical overhead stirrer. The reactions were carried out at the selected temperature for the chosen time. At the end of each reaction, the reactors were rapidly cooled at room temperature by blown air, the hydrolyze was separated from solid residuals by vacuum filtration, filtered with a PTFE filter (0.2 µm) and analyzed through high-pressure liquid chromatography (HPLC) with a refractive index detector.

3.5. Acid-Catalyzed Alcoholysis of Cardoon

Alcoholysis reactions were carried out in n-butanol, using H_2SO_4 as catalyst, in the MW reactor CEM Discover S-class System at 190 °C for the selected reaction time. At the end of each reaction, the slurry was filtered under vacuum on a crucible and the liquid samples were diluted with acetone and analyzed by a gas chromatograph coupled with a flame ionization detector (GC-FID).

3.6. Analytical Techniques

Liquid samples deriving from hydrolysis were analyzed by an HPLC system (Perkin Elmer Flexer Isocratic Platform) equipped with a Benson 2000-0 BP-OA column (300 mm

× 7.8 mm) and coupled with a Waters 2140 refractive index detector. A 0.005 M H_2SO_4 aqueous solution was adopted as mobile phase, maintaining the column at 60 °C with the flow rate of 0.6 mL/min. The concentrations of hydrolysis products were determined from calibration curves obtained with external standard solutions. Each analysis was carried out in duplicate and the reproducibility of this analysis was within 3%.

Liquid samples from alcoholysis were analyzed by a GC-FID instrument (DANI GC1000 DPC) equipped with a fused silica capillary column–HP-PONA cross-linked methyl silicone gum (20 m × 0.2 mm × 0.5 µm). The injection and flame ionization detector ports were set at 250 °C. The oven temperature program was set at 90 °C for 3 min and then increased at the rate of 10 °C/min till it reached 260 °C, where it was maintained for 5 min and up to 280 °C with the rate of 10 °C/min and maintained for 3 min. Nitrogen was used as the carrier gas, at the flow rate of 0.2 mL/min. The quantitative analysis of alcoholysis products was performed by calibration using n-dodecane as internal standard. Each analysis was carried out in duplicate and the reproducibility of this analysis was within 5%.

The molar and the ponderal yields of the compounds of interest were calculated according to the following equations, respectively:

Molar Yield (mol%) = [product (mol)/anhydrous glucose unit in starting biomass (mol)] × 100;

Ponderal Yield (wt%) [product (g)/dry starting biomass (g)] × 100.

Analysis of impurities was performed by GC-MS and HPLC-MS. When GC-MS analysis was carried out, the starting aqueous hydrolysate was extracted with diethyl ether and the diluted extract (about 0.1–0.2 µL) was qualitatively analyzed by the instrument Hewlett-Packard HP 7890 (Palo Alto, CA, USA), equipped with an MSDHP 5977 detector and with a G.C. column Phenomenex Zebron with a 100% methyl polysiloxane stationary phase (column length 30 m, inner diameter 0.25 mm and thickness of the stationary phase 0.25 µm), in splitless mode. The transport gas was helium 5.5 and the flow was 1 mL/min. The temperature of the injection port was set at 250 °C, carrier pressure at 100 kPa. The oven was heated at 50 °C for 1 min, then the temperature was raised at 3 °C/min up to 250 °C and held for 5 min. When HPLC-MS analysis was carried out, a Sciex X500 qTOF mass spectrometer (Sciex, Darmstadt, Germany) was coupled to an HPLC Agilent 1260 Infinity II (Agilent, Waldbronn, Germany) and operated in ESI negative mode (spray voltage of −4500 V). HPLC system was equipped with a diode-array detector (DAD), which operated at 210, 250 and 280 nm. Chromatographic separation was achieved on a Zorbax SB-C18 column 150 mm × 4.6 mm, particle size 3.5 µm (Agilent, California, United States) as the stationary phase, using water with 0.1% (v/v) of formic acid (A) and methanol (B) as the eluents. The mobile phase flow rate was 0.7 mL/min, and the column oven temperature was set at 25 °C. The gradient was programmed as follows: 0–22.0 min at 95% (A), 22.0–30.0 min at 5% (A), 30.0 min at 95% (A), held for 5 min. The sample was filtered and 10 µL were injected into the HPLC system, after proper dilution (1:1) and filtration.

Fourier transform infrared (FT-IR) spectra for raw biomasses and solid residues after reactions were recorded in attenuated total reflection (ATR) mode with a Spectrum-Two Perkin-Elmer spectrophotometer. The acquisition of each spectrum provided 12 scans, with a resolution of 8 cm^{-1}, in the wavenumber range between 4000 and 450 cm^{-1}.

Thermogravimetric analysis (TGA) of starting feedstocks and solid residues after reactions was performed with a Mettler Toledo TGA/SDTA 851 apparatus in high purity N_2. The sample was heated from 30 °C up to 900 °C, at a rate of 10 °C/min, under nitrogen atmosphere (60 mL/min). Both weight loss and weight loss rate were acquired during each experiment.

Elemental analysis (C, H, N, S) of starting feedstocks and solid residues after reactions was performed by a commercially available automatic analyzer Elementar Vario MICRO Cube (Elementar, Germany). These elements were quantified adopting a thermal conductivity detector (TCD). Lastly, oxygen content was calculated by difference: O (%) = 100

(%)–C (%)–H (%)–N (%)–S (%). HHV was calculated according to the following correlation proposed by Channiwala and Parikh [63]:

HHV (MJ/Kg) = 0.3491 C(%) + 1.1783 H(%) + 0.1005 S(%) − 0.1034 O(%) − 0.0151N (%) − 0.0211 Ash(%).

3.7. Purification of the Crude Mother Liquor

Crude mother liquor was extracted by continuous liquid/liquid extraction of the aqueous mixture with 2-methyltetrahydrofuran as the extraction solvent. In a typical procedure, 20 mL of the crude hydrolyzate was treated with 60 mL of 2-methyltetrahydrofuran in a continuous liquid/liquid extractor apparatus for 4 h and then, once the organic fraction was separated, the extract was subjected to a fractional distillation, in order to remove the solvent and recover the LA as pure product, according to the general procedure previously patented [64]. Briefly, the extract was subjected to a first distillation step, in order to remove the solvent, working under atmospheric pressure, until the bottom temperature was 100 °C. The bottom product was further distilled, in order to remove any lights, using a 50 cm Vigreux column, working at 100 mbar, progressively increasing the temperature of the oil bath from 75 to 130 °C (corresponding temperatures of the bottom zone of the distiller in the range 40–115 °C), and the distillation was stopped when no vapors reached the top of the distillation apparatus. Lastly, the bottom product was further distilled, working at 5 mbar, increasing the oil bath temperature from 185 to 195 °C (corresponding temperatures of the bottom zone of the distiller in the range 135–145 °C). The isolated top fraction was dried by a mechanical pump and the dried fraction was characterized by HPLC and GC chromatography. The LA purity grade of 93% was ascertained and on the basis of weighted amount and taking into account the LA purity grade, the isolated yield was determined.

4. Conclusions

In this paper, the acid-catalyzed hydrolysis and alcoholysis of waste cardoon residues from agroindustry to give strategic platforms LA and BL, respectively, were investigated. After removal of oil seeds, the cardoon was employed directly and after a steam-explosion pretreatment, this last affording a cellulose-rich feedstock. This positive effect of the SE pretreatment can be applied to a larger scale only if its economic sustainability is verified: in this sense a specific cost-benefit analysis should be developed case by case, considering that the more expensive subsection of the SE facility is the vapor generator.

MW-assisted hydrolysis reactions on this last biomass led to high values of LA molar yield and concentration, up to 49.9 mol% and 62.1 g/L, respectively. The hydrolysis was also performed in a traditional batch autoclave, reaching LA yield and concentration of 47.1 mol% and 41.4 g/L, respectively, demonstrating that such a process can be successfully intensified, by switching to traditional industrial reactors. The achieved results are really interesting, highlighting that eco-friendly reaction conditions, such as high biomass to acid catalyst ratio, water as the reactant/reaction medium and energy-saving heating, can be adopted for the conversion of low-cost residual cardoon, implying that sustainable exploitation of such biomass can be developed. Moreover, a preliminary study on the acid-catalyzed butanolysis of steam-exploded cardoon led to good BL yields, up to 42.5 mol%, proving that the direct one-pot conversion of the cellulosic fraction of this waste biomass can supply a wider range of value-added products, including bio-fuels. Moreover, the characterization of the solid residues recovered from both processes has allowed us to envisage their possible exploitation, in a perspective of the complete valorization of cardoon biomass.

The proposed approaches represent alternative solutions to reduce consumption of fossil resources and carbon dioxide emissions and recycle massive amounts of agricultural residues, in agreement with the concept of third-generation biorefinery.

Supplementary Materials: The following are available online at https://www.mdpi.com/article/10.3390/catal11091082/s1, Figure S1: Total ion chromatogram (TIC) of the diethyl ether extract and corresponding mass spectra of the identified compounds, Figure S2: RI-and UV-HPLC chromatogram of the crude hydrolysate at 284 nm, Figure S3: UV-HPLC chromatograms of the crude hydrolysate at 205 and 284 nm, Figure S4: TIC related to all ions of all detected masses and corresponding UV chromatograms (280, 250 nm, superimposed) and 205 nm, Figure S5: XIC chromatogram obtained from selected compounds of interest, according to Glińska et al. For numbering and corresponding assignments, see Table S2, Figure S6: XIC chromatogram obtained from selected compounds of interest, according to Ibáñez et al. For numbering and corresponding assignments, see Table S3, Figure S7: Monomeric/oligomeric (as dimeric/trimeric) compounds of 5-HMF and corresponding XIC chromatogram, Figure S8 and Table S1: Chromatographic data of the main detected compounds reported in Figure S4, Table S2: Chromatographic data of selected compounds of interest for hydrolyzates of biomass source, according to Glińska et al. (see also XIC chromatogram of Figure S5), Table S3: Chromatographic data of selected compounds of interest for hydrolyzates of biomass source, according to Ibáñez et al. (see also XIC chromatogram of Figure S6).

Author Contributions: A.M.R.G., C.A., S.C., and D.L. conceived the experiments; A.M.R.G., S.C., D.L., N.D.F. and C.A. designed the experiments; N.D.F., S.C., V.C. and D.L. performed the experiments and analysis; all the authors analyzed the data; A.M.R.G., C.A., S.C., V.C. and D.L. wrote the paper; N.D.F. and F.C. revised and supervised the writing of the manuscript. All authors have read and agreed to the published version of the manuscript.

Funding: This research was funded by the project VISION PRIN 2017 FWC3WC_002 funded by MIUR and by the project PRA_2018_26 funded by University of Pisa.

Acknowledgments: The project VISION PRIN 2017 FWC3WC_002 and the project PRA_2018_26 of University of Pisa are gratefully acknowledged.

Conflicts of Interest: The authors declare no conflict of interest. The funders had no role in the design of the study; in the collection, analyses, or interpretation of data; in the writing of the manuscript, or in the decision to publish the results.

References

1. Puricelli, S.; Cardellini, G.; Casadei, S.; Faedo, D.; Van den Oever, A.E.M.; Grosso, M. A review on biofuels for light-duty vehicles in Europe. *Renew. Sustain. Energy Rev.* **2021**, *137*, 110398. [CrossRef]
2. Falcone, P.M.; Lopolito, A.; Sica, E. The networking dynamics of the Italian biofuel industry in time of crisis: Finding an effective instrument mix for fostering a sustainable energy transition. *Energy Policy* **2018**, *112*, 334–348. [CrossRef]
3. US Environmental Protection Agency. Overview of Greenhouse Gases. Available online: https://www.epa.gov/ghgemissions/overview-greenhouse-gases (accessed on 24 June 2021).
4. Intergovernmental Panel on Climate Change. *Climate Change 2014: Synthesis Report*; Contribution of Working Groups I, II and III to the Fifth Assessment Report of the Intergovernmental Panel on Climate Change; Core Writing Team; Pachauri, R.K., Meyer, L.A., Eds.; Intergovernmental Panel on Climate Change: Geneva, Switzerland, 2014; p. 151.
5. Coccia, V.; Cotana, F.; Cavalaglio, G.; Gelosia, M.; Petrozzi, A. Cellulose Nanocrystals Obtained from Cynara Cardunculus and Their Application in the Paper Industry. *Sustainability* **2014**, *6*, 5252. [CrossRef]
6. Barracosa, P.; Barracosa, M.; Pires, E. Cardoon as a Sustainable Crop for Biomass and Bioactive Compounds Production. *Chem. Biodivers.* **2019**, *16*, e1900498. [CrossRef]
7. Zayed, A.; Serag, A.; Farag, M.A. Cynara cardunculus L.: Outgoing and potential trends of phytochemical, industrial, nutritive and medicinal merits. *J. Funct. Foods* **2020**, *69*, 103937. [CrossRef]
8. Grammelis, P.; Malliopoulou, A.; Basinas, P.; Danalatos, N.G. Cultivation and Characterization of Cynara Cardunculus for Solid Biofuels Production in the Mediterranean Region. *Int. J. Mol. Sci.* **2008**, *9*, 1241. [CrossRef]
9. Fernández, J.; Curt, M.D.; Aguado, P.L. Industrial applications of Cynara cardunculus L. for energy and other uses. *Ind. Crop. Prod.* **2006**, *24*, 222–229. [CrossRef]
10. Espada, J.J.; Villalobos, H.; Rodríguez, R. Environmental assessment of different technologies for bioethanol production from Cynara cardunculus: A Life Cycle Assessment study. *Biomass Bioenergy* **2021**, *144*, 105910. [CrossRef]
11. Sun, Y.; Cheng, J. Hydrolysis of lignocellulosic materials for ethanol production: A review. *Bioresour. Technol.* **2002**, *83*, 1–11. [CrossRef]
12. Overend, R.P.; Chornet, E. Fractionation of lignocellulosics by steam-aqueous pretreatments. *Philos. Trans. R. Soc. Lond. Ser. A Math. Phys. Sci.* **1987**, *321*, 526–533. [CrossRef]
13. Antonetti, C.; Licursi, D.; Fulignati, S.; Valentini, G.; Raspolli Galletti, A.M. New Frontiers in the Catalytic Synthesis of Levulinic Acid: From Sugars to Raw and Waste Biomass as Starting Feedstock. *Catalysts* **2016**, *6*, 196. [CrossRef]

14. Bozell, J.J.; Petersen, G.R. Technology development for the production of biobased products from biorefinery carbohydrates—the US Department of Energy's "Top 10" revisited. *Green Chem.* **2010**, *12*, 539–554. [CrossRef]
15. Leal Silva, J.F.; Grekin, R.; Mariano, A.P.; Maciel Filho, R. Making Levulinic Acid and Ethyl Levulinate Economically Viable: A Worldwide Technoeconomic and Environmental Assessment of Possible Routes. *Energy Technol.* **2018**, *6*, 613–639. [CrossRef]
16. Christensen, E.; Williams, A.; Paul, S.; Burton, S.; McCormick, R. Properties and Performance of Levulinate Esters as Diesel Blend Components. *Energy Fuels* **2011**, *25*, 5422–5428. [CrossRef]
17. Howard, M.S.; Issayev, G.; Naser, N.; Sarathy, S.M.; Farooq, A.; Dooley, S. Ethanolic gasoline, a lignocellulosic advanced biofuel. *Sustain. Energy Fuels* **2019**, *3*, 409–421. [CrossRef]
18. Licursi, D.; Antonetti, C.; Fulignati, S.; Giannoni, M.; Raspolli Galletti, A.M. Cascade Strategy for the Tunable Catalytic Valorization of Levulinic Acid and γ-Valerolactone to 2-Methyltetrahydrofuran and Alcohols. *Catalysts* **2018**, *8*, 277. [CrossRef]
19. Rivas, S.; Raspolli Galletti, A.M.; Antonetti, C.; Licursi, D.; Santos, V.; Parajó, J.C. A Biorefinery Cascade Conversion of Hemicellulose-Free Eucalyptus Globulus Wood: Production of Concentrated Levulinic Acid Solutions for γ-Valerolactone Sustainable Preparation. *Catal.* **2018**, *8*, 169. [CrossRef]
20. Rackemann, D.W.; Doherty, W. The conversion of lignocellulosics to levulinic acid. *Biofuels Bioprod. Biorefining* **2011**, *5*, 198–214. [CrossRef]
21. Timokhin, B.V.; A Baransky, V.; Eliseeva, G.D. Levulinic acid in organic synthesis. *Russ. Chem. Rev.* **1999**, *68*, 73–84. [CrossRef]
22. Hayes, D.J.; Fitzpatrick, S.; Hayes, M.H.B.; Ross, J.R.H. The Biofine Process–Production of Levulinic Acid, Furfural, and Formic Acid from Lignocellulosic Feedstocks. In *Biorefineries–Industrial Processes and Product: Status Quo and Future Directions*; Kamm, B., Gruber, P.R., Kamm, M., Eds.; Wiley-VCH: Weinheim, Germany, 2006; Volume 1, pp. 139–164.
23. Morone, A.; Apte, M.; Pandey, R.A. Levulinic acid production from renewable waste resources: Bottlenecks, potential remedies, advancements and applications. *Renew. Sustain. Energy Rev.* **2015**, *51*, 548–565. [CrossRef]
24. Démolis, A.; Essayem, N.; Rataboul, F. Synthesis and Applications of Alkyl Levulinates. *ACS Sustain. Chem. Eng.* **2014**, *2*, 1338–1352. [CrossRef]
25. Hishikawa, Y.; Yamaguchi, M.; Kubo, S.; Yamada, T. Direct preparation of butyl levulinate by a single solvolysis process of cellulose. *J. Wood Sci.* **2013**, *59*, 179–182. [CrossRef]
26. Démolis, A.; Eternot, M.; Essayem, N.; Rataboul, F. Influence of butanol isomers on the reactivity of cellulose towards the synthesis of butyl levulinates catalyzed by liquid and solid acid catalysts. *New J. Chem.* **2016**, *40*, 3747–3754. [CrossRef]
27. Antonetti, C.; Gori, S.; Licursi, D.; Pasini, G.; Frigo, S.; López, M.; Parajó, J.C.; Raspolli Galletti, A.M. One-Pot Alcoholysis of the Lignocellulosic Eucalyptus nitens Biomass to n-Butyl Levulinate, a Valuable Additive for Diesel Motor Fuel. *Catalysis* **2020**, *10*, 509. [CrossRef]
28. Matrìca. Green Chemicals. Available online: https://www.matrica.it/Default.asp?ver=en (accessed on 29 June 2021).
29. Mosier, N.; Wyman, C.; Dale, B.; Elander, R.; Lee, Y.Y.; Holtzapple, M.; Ladisch, M. Features of promising technologies for pretreatment of lignocellulosic biomass. *Bioresour. Technol.* **2005**, *96*, 673–686. [CrossRef] [PubMed]
30. Cotana, F.; Cavalaglio, G.; Gelosia, M.; Coccia, V.; Petrozzi, A.; Nicolini, A. Effect of Double-Step Steam Explosion Pretreatment in Bioethanol Production from Softwood. *Appl. Biochem. Biotechnol.* **2014**, *174*, 156–167. [CrossRef]
31. Novamont. *(BIT3G) Project—Third Generation Biorefinery Spread on the Territory, Supported by the Italian Ministry of Education (MIUR) and Coordinated by Novamont*; Novamont: Novara, Italy, 2014.
32. Asghari, F.S.; Yoshida, H. Acid-Catalyzed Production of 5-Hydroxymethyl Furfural from d-Fructose in Subcritical Water. *Ind. Eng. Chem. Res.* **2006**, *45*, 2163–2173. [CrossRef]
33. Flannelly, T.; Lopes, M.; Kupiainen, L.; Dooley, S.; Leahy, J.J. Non-stoichiometric formation of formic and levulinic acids from the hydrolysis of biomass derived hexose carbohydrates. *RSC Adv.* **2015**, *6*, 5797–5804. [CrossRef]
34. Shi, N.; Liu, Q.; Ju, R.; He, X.; Zhang, Y.; Tang, S.; Ma, L. Condensation of α-Carbonyl Aldehydes Leads to the Formation of Solid Humins during the Hydrothermal Degradation of Carbohydrates. *ACS Omega* **2019**, *4*, 7330–7343. [CrossRef]
35. Martinez, A.; Rodriguez, M.E.; York, S.W.; Preston, J.F.; Ingram, L.O. Use of UV Absorbance To Monitor Furans in Dilute Acid Hydrolysates of Biomass. *Biotechnol. Prog.* **2000**, *16*, 637–641. [CrossRef]
36. Glińska, K.; Lerigoleur, C.; Giralt, J.; Torrens, E.; Bengoa, C. Valorization of Cellulose Recovered from WWTP Sludge to Added Value Levulinic Acid with a Brønsted Acidic Ionic Liquid. *Catalysts* **2020**, *10*, 1004. [CrossRef]
37. Ibáñez, A.B.; Bauer, S. Analytical method for the determination of organic acids in dilute acid pretreated biomass hydrolysate by liquid chromatography-time-of-flight mass spectrometry. *Biotechnol. Biofuels* **2014**, *7*, 145–153. [CrossRef]
38. Iroba, K.L.; Tabil, L.G.; Sokhansanj, S.; Dumonceaux, T. Pretreatment and fractionation of barley straw using steam explosion at low severity factor. *Biomass Bioenergy* **2014**, *66*, 286–300. [CrossRef]
39. Schimmelpfennig, S.; Glaser, B. One Step Forward toward Characterization: Some Important Material Properties to Distinguish Biochars. *J. Environ. Qual.* **2012**, *41*, 1001–1013. [CrossRef]
40. Licursi, D.; Antonetti, C.; Bernardini, J.; Cinelli, P.; Coltelli, M.B.; Lazzeri, A.; Martinelli, M.; Raspolli Galletti, A.M. Characterization of the Arundo Donax L. solid residue from hydrothermal conversion: Comparison with technical lignins and application perspectives. *Ind. Crop. Prod.* **2015**, *76*, 1008–1024. [CrossRef]
41. Licursi, D.; Antonetti, C.; Fulignati, S.; Vitolo, S.; Puccini, M.; Ribechini, E.; Bernazzani, L.; Raspolli Galletti, A.M. In-depth characterization of valuable char obtained from hydrothermal conversion of hazelnut shells to levulinic acid. *Bioresour. Technol.* **2017**, *244*, 880–888. [CrossRef]

42. Düdder, H.; Wütscher, A.; Stoll, R.; Muhler, M. Synthesis and characterization of lignite-like fuels obtained by hydrothermal carbonization of cellulose. *Fuel* **2016**, *171*, 54–58. [CrossRef]
43. Di Fidio, N.; Raspolli Galletti, A.M.; Fulignati, S.; Licursi, D.; Liuzzi, F.; De Bari, I.; Antonetti, C. Multi-Step Exploitation of Raw *Arundo donax* L. for the Selective Synthesis of Second-Generation Sugars by Chemical and Biological Route. *Catalysts* **2020**, *10*, 79. [CrossRef]
44. Fiore, V.; Scalici, T.; Valenza, A. Characterization of a new natural fiber from Arundo donax L. as potential reinforcement of polymer composites. *Carbohydr. Polym.* **2014**, *106*, 77–83. [CrossRef] [PubMed]
45. Liu, H.-M.; Li, H.-Y.; Li, M.-F. Cornstalk liquefaction in sub- and super-critical ethanol: Characterization of solid residue and the liquefaction mechanism. *J. Energy Inst.* **2017**, *90*, 734–742. [CrossRef]
46. Tsilomelekis, G.; Orella, M.J.; Lin, Z.; Cheng, Z.; Zheng, W.; Nikolakis, V.; Vlachos, D.G. Molecular structure, morphology and growth mechanisms and rates of 5-hydroxymethyl furfural (HMF) derived humins. *Green Chem.* **2016**, *18*, 1983–1993. [CrossRef]
47. Van Zandvoort, I.; Wang, Y.; Rasrendra, C.B.; Van Eck, E.R.H.; Bruijnincx, P.C.A.; Heeres, H.J.; Weckhuysen, B.M. Formation, Molecular Structure, and Morphology of Humins in Biomass Conversion: Influence of Feedstock and Processing Conditions. *ChemSusChem* **2013**, *6*, 1745–1758. [CrossRef]
48. Dee, S.J.; Bell, A.T. A Study of the Acid-Catalyzed Hydrolysis of Cellulose Dissolved in Ionic Liquids and the Factors Influencing the Dehydration of Glucose and the Formation of Humins. *ChemSusChem* **2011**, *4*, 1166–1173. [CrossRef] [PubMed]
49. Antonetti, C.; Raspolli Galletti, A.M.; Fulignati, S.; Licursi, D. Amberlyst A-70: A surprisingly active catalyst for the MW-assisted dehydration of fructose and inulin to HMF in water. *Catal. Commun.* **2017**, *97*, 146–150. [CrossRef]
50. Antonetti, C.; Fulignati, S.; Licursi, D.; Raspolli Galletti, A.M. Turning Point toward the Sustainable Production of 5-Hydroxymethyl-2-furaldehyde in Water: Metal Salts for Its Synthesis from Fructose and Inulin. *ACS Sustain. Chem. Eng.* **2019**, *7*, 6830–6838. [CrossRef]
51. Nasir, N.; Davies, G.; McGregor, J. Tailoring product characteristics in the carbonisation of brewers' spent grain through solvent selection. *Food Bioprod. Process.* **2019**, *120*, 41–47. [CrossRef]
52. Díez, D.; Urueña, A.; Piñero, R.; Barrio, A.; Tamminen, T. Determination of Hemicellulose, Cellulose, and Lignin Content in Different Types of Biomasses by Thermogravimetric Analysis and Pseudocomponent Kinetic Model (TGA-PKM Method). *Processes* **2020**, *8*, 1048. [CrossRef]
53. Licursi, D.; Antonetti, C.; Mattonai, M.; Pérez-Armada, L.; Rivas, S.; Ribechini, E.; Raspolli Galletti, A.M. Multi-valorisation of giant reed (*Arundo Donax* L.) to give levulinic acid and valuable phenolic antioxidants. *Ind. Crop. Prod.* **2018**, *112*, 6–17. [CrossRef]
54. Yu, L.; Falco, C.; Weber, J.; White, R.J.; Howe, J.Y.; Titirici, M.-M. Carbohydrate-Derived Hydrothermal Carbons: A Thorough Characterization Study. *Langmuir* **2012**, *28*, 12373–12383. [CrossRef] [PubMed]
55. Rasrendra, C.; Windt, M.; Wang, Y.; Adisasmito, S.; Makertihartha, I.G.B.N.; van Eck, E.R.H.; Meier, D.; Heeres, H.J. Experimental studies on the pyrolysis of humins from the acid-catalysed dehydration of C6-sugars. *J. Anal. Appl. Pyrolysis* **2013**, *104*, 299–307. [CrossRef]
56. Antonetti, C.; Licursi, D.; Raspolli Galletti, A.M. New Intensification Strategies for the Direct Conversion of Real Biomass into Platform and Fine Chemicals: What Are the Main Improvable Key Aspects? *Catalysts* **2020**, *10*, 961. [CrossRef]
57. Jindo, K.; Sánchez-Monedero, M.A.; Mastrolonardo, G.; Audette, Y.; Higashikawa, F.S.; Silva, C.A.; Akashi, K.; Mondini, C. Role of biochar in promoting circular economy in the agriculture sector. Part 2: A review of the biochar roles in growing media, composting and as soil amendment. *Chem. Biol. Technol. Agric.* **2020**, *7*, 16–25. [CrossRef]
58. Liu, W.-J.; Jiang, H.; Yu, H.-Q. Development of Biochar-Based Functional Materials: Toward a Sustainable Platform Carbon Material. *Chem. Rev.* **2015**, *115*, 12251–12285. [CrossRef]
59. Sluiter, A.; Hames, B.; Ruiz, R.; Scarlata, C.; Sluiter, J.; Templeton, D.; Crocker, D. *Determination of Structural Carbohydrates and Lignin in Biomass*. NREL/TP-510-42618; Scarlata, C., Sluiter, J., Templeton, D., Eds.; National Renewable Energy Laboratory: Golden, CO, USA, 2008.
60. Sluiter, A.; Ruiz, R.; Scarlata, C.; Sluiter, J.; Templeton, D. *Determination of Extractives in Biomass*. NREL/TP-510-42619; Sluiter, J., Ed.; National Renewable Energy Laboratory: Golden, CO, USA, 2008.
61. Sluiter, A.; Hames, B.; Ruiz, R.; Scarlata, C.; Sluiter, J.; Templeton, D. *Determination of Ash in Biomass.*, NREL/TP-510-42622; National Renewable Energy Laboratory: Golden, CO, USA, 2008.
62. Sluiter, A.; Hames, B.; Hyman, D.; Payne, C.; Ruiz, R.; Scarlata, C.; Sluiter, J.; Templeton, D.; Wolfe, J. *Determination of Total Solids in Biomass and Total Dissolved Solids in Liquid Process Samples*. NREL/TP-510-42621; National Renewable Energy Laboratory: Golden, CO, USA, 2008.
63. Channiwala, S.A.; Parikh, P.P. A unified correlation for estimating HHV of solid, liquid and gaseous fuels. *Fuel* **2002**, *81*, 1051–1063. [CrossRef]
64. Woestenborghs, P.L.; Altink, R.L. Process for the Isolation of Levulinic Acid. Patent No WO 2015/007602 A1, 22 January 2015.

Article

Hydrogenative Cyclization of Levulinic Acid to γ-Valerolactone with Methanol and Ni-Fe Bimetallic Catalysts

Ligang Luo [1,*,†], **Xiao Han** [1,†] **and Qin Zeng** [1,2]

1. College of Life Science, Shanghai Normal University, 100 Guilin Rd, Shanghai 200241, China; wo.lixiang0530@163.com (X.H.); zengqllg@usst.edu.cn (Q.Z.)
2. The Public Experiment Center, University of Shanghai for Science and Technology, Shanghai 200093, China
* Correspondence: liangluo@shnu.edu.cn
† Ligang Luo and Xiao Han Contributed equally to this work.

Received: 26 August 2020; Accepted: 14 September 2020; Published: 21 September 2020

Abstract: A series of Ni-Fe/SBA-15 catalysts was prepared and tested for the catalytic hydrogenation of levulinic acid to γ-valerolactone, adopting methanol as the only hydrogen donor, and investigating the synergism between Fe and Ni, both supported on SBA-15, towards this reaction. The characterization of the synthesized catalysts was carried out by XRD (X-ray powder diffraction), TEM (transmission electron microscopy), H_2-TPD (hydrogen temperature-programmed desorption), XPS (X-ray photoelectron spectroscopy), and in situ FT-IR (Fourier transform–infrared spectroscopy) techniques. H_2-TPD and XPS results have shown that electron transfer occurs from Fe to Ni, which is helpful both for the activation of the C=O bond and for the dissociative activation of H_2 molecules, also in agreement with the results of the in situ FT-IR spectroscopy. The effect of temperature and reaction time on γ-valerolactone production was also investigated, identifying the best reaction conditions at 200 °C and 180 min, allowing for the complete conversion of levulinic acid and the complete selectivity to γ-valerolactone. Moreover, methanol was identified as an efficient hydrogen donor, if used in combination with the Ni-Fe/SBA-15 catalyst. The obtained results are promising, especially if compared with those obtained with the traditional and more expensive molecular hydrogen and noble-based catalysts.

Keywords: biomass ester derivatives; solvothermal processing; levulinic acid; γ-valerolactone; Ni-Fe bimetallic catalysts

---

1. Introduction

The world is highly dependent on the utilization of fossil resources to fulfill energy needs for the production of heat and power. Nevertheless, the surplus consumption of fossil fuels is escalating the concentration of atmospheric CO_2, which causes serious global warming threats [1,2]. As a solution to the above primary challenges, considerable attention has been paid to the exploitation of renewable resources [1]. In this context, the conversion of the cellulose fraction of biomass into value-added platform chemicals, such as 5-hydroxymethylfurfural (5-HMF) and levulinic acid (LA), has been thoroughly investigated [2,3]. Moreover, the transformation of LA into more added-value bio-fuels and bio-chemicals is strategic, thanks to its reactive keto and carboxylic functional groups, which can be exploited for the synthesis of many valuable bio-chemicals, such as fuel additives, fragrances, solvents, pharmaceuticals, and plasticizers [4]. Among these high-value chemicals, γ-valerolactone (GVL) is receiving considerable attention to synthesize added-value bio-chemicals, such as food additives, solvents, and drug intermediates, as well as new bio-fuels [5,6]. Three pathways have been proposed for GVL production, as reported in Scheme 1. The first requires the high-temperature LA conversion in the gas phase, occurring via its endothermic dehydration to angelicalactone, which is subsequently hydrogenated to GVL. The second occurs in the liquid phase, at lower temperatures, where the LA keto group is reduced to 4-hydroxyvaleric acid, which is in turn dehydrated to GVL. Lastly, the third possibility provides the esterification of LA to the less reactive alkyl levulinate, which undergoes hydrogenation and dehydration to GVL [6].

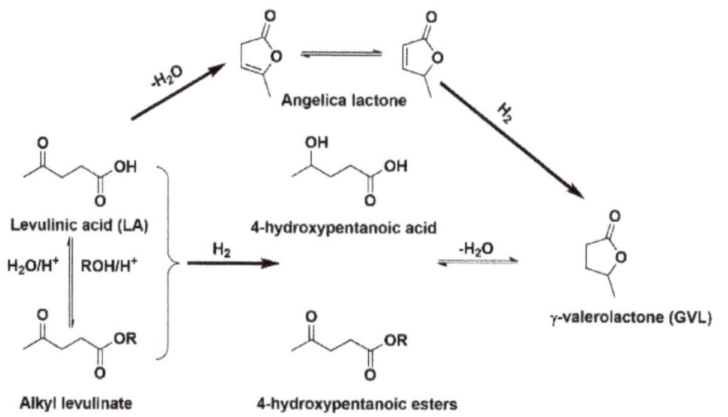

Scheme 1. Different pathways for the hydrogenation of levulinic acid (LA) to γ-valerolactone (GVL).

According to the above mechanisms, dehydration is generally favored by the presence of acid catalysts, whereas the hydrogenation is catalyzed by transition metals. Many homogeneous and heterogeneous catalysts have been employed for the hydrogenation of LA to GVL in the liquid phase [7]. However, given the well-known drawbacks of the homogeneous catalysts (the possible corrosion of the equipment, difficult recycle/reuse, and environmental concerns), heterogeneous ones have been certainly preferred [8]. Many noble metal catalysts have been tested for this reaction, in particular, Pt, Au, Pd, Rh, and Ru, in many cases achieving promising catalytic performances with respect to GVL [9–12]. For example, Son et al. have studied the catalytic performances of Ru/C, Ru/SBA, Au/ZrC, and Au/ZrO$_2$ catalysts for the LA hydrogenation, highlighting the good catalytic performances and stability of the Au/ZrO$_2$ catalyst, which can be reused five times, achieving a high GVL yield (90 mol %) [10]. However, although the noble metal catalysts show excellent performances for the hydrogenation of such carbonyl compounds, the high costs limit their further application in the industry. On this basis, non-noble metal catalysts remain the preferred choice for improving the sustainable catalytic transformation of LA to GVL [13,14]. In particular, Ni-based catalysts (Raney-Ni, Ni/MoOx/C, Cu/ZrO$_2$, NiCu/Al$_2$O$_3$, and Mo$_2$C) are active in this reaction [13–15]. For example, Mohan et al. have examined the influence of catalysts with Ni supported on SiO$_2$, Al$_2$O$_3$, ZnO, ZrO$_2$, TiO$_2$, and MgO, highlighting the excellent catalytic performances of Ni/SiO$_2$ towards the LA hydrogenation to GVL (0.8506 kg GVL kg catalyst^{-1} h^{-1} at 250 °C) [15]. The characterization of these catalysts by FT-IR spectroscopy has indicated that both Lewis and Brönsted acid sites play an important role in the dehydration of the intermediate 4-hydroxypentanoic acid [16], thus highlighting that the use of tunable bifunctional catalysts is preferred for improving the catalysis of this reaction. In this context, the addition of another metal could modify the electronic structure of Ni, decreasing the interaction between active metal and supports, thus enhancing the reaction activity [17]. For example, the Cu species had an important effect on the chemical environment of Ni species and acted as an electronic promoter on Ni surface active sites in Ni-Cu/Al$_2$O$_3$ bimetallic catalysts, achieving the highest GVL yield of 96 mol % (250 °C, 6.5 MPa, and 2 h) [18]. Regarding the possible hydrogen sources, LA hydrogenation has been carried out mainly with molecular H$_2$ (0.8–6.5 MPa), both in batch autoclaves and in continuous flow reactors, generally obtaining high yields and selectivities to GVL [15–17]. However, this reaction requires large quantities of molecular H$_2$, whose production is still mostly fossil based, which also raises cost and safety issues. In this context, formic acid and alcohols have been proposed as alternative hydrogen donors, which should help one to avoid the usage of external H_2 and, subsequently, the requirement for a high-pressure reactor [19]. Regarding formic acid, H_2 is generated through the decarboxylation route (HCOOH → CO_2 + H_2), which is generally slow, unless noble metals, such as Ru, are used as catalysts [19,20]. However, Ruppert et al. have demonstrated that formic acid can be irreversibly adsorbed on the Ru/C catalyst, hampering the LA conversion to GVL [21]. Furthermore, agglomeration and metal leaching generally occur, which could greatly affect catalyst stability, thus significantly hampering its effective use [22]. Alternatively, it is possible to use alcohol as an H_2 source, and perform a hydrogen transfer, which is known as the Meerwein–Ponndorf–Verley (MPV) reaction, which is catalyzed by Lewis acid sites, such as those of metal oxides or zeolites [23]. Two mechanisms have been proposed for explaining the hydrogen transfer in the presence of alcohols, both reported in Scheme 2 [23–25]. In the first one (path A, Scheme 2), the MPV reaction is catalyzed by the metal catalyst, which receives the hydrogen from the alcohol donor, through β-hydride elimination, and the activated hydrogen is then transferred from the metal to the acceptor molecule, in this case LA [24]. The second one involves an intermolecular hydrogen transfer, which occurs by adsorption of the hydroxyl group of the alcohol and the carbonyl group of LA (or its ester) on the catalyst surface [25]. This direct reaction is allowed thanks to the presence of Lewis and Brönsted acids, both leading to the formation of a six-membered ring transition state. Luo et al. have studied the MPV reaction of methyl levulinate to GVL in the presence of zeolites, confirming that the main rate-limiting step of this reaction is the hydride shift [25].

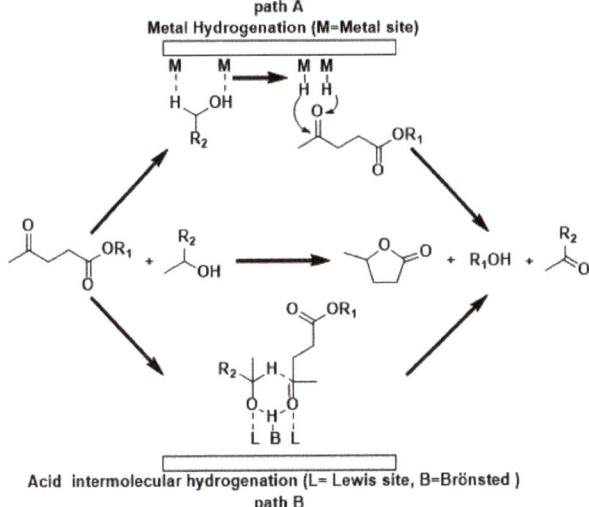

Scheme 2. Catalytic transfer hydrogenation mechanisms of LA.

Many kinds of alcohol have been employed for the MPV reaction, highlighting the higher reaction rates of the secondary alcohols. Chia et al. employed 2-BuOH to study the MPV transfer hydrogenation, achieving a good GVL yield (84.7 mol %), over the ZrO_2 catalyst (150 °C, 16 h) [24]. A strong correlation has been established between the reducing capacity of alcohols and MPV reduction, according to this order: MeOH < EtOH < 1-BuOH < 2-BuOH = 2-PrOH [26]. However, by using MeOH or EtOH as hydrogen donors, more efficient and integrated biomass utilization chains could be created, thanks to the high availability of these alcohols from biomass feedstock. Tang et al. have used EtOH as a hydrogen donor, achieving the maximum GVL yield of 64 mol % (250 °C, 3 h) [26]. Yi et al. have reported that no hydrogen transfer observed within 48 h in the presence of EtOH, whilst a GVL yield of 38 mol % was achieved with 1 MPa H_2, after only 2 h [27]. Despite the achieved progress, it is definitely still necessary to develop new, efficient and cheap catalytic systems for the LA hydrogenation to GVL, by MPV catalytic transfer, in the presence of short-chain primary alcohols.

In this work, new bimetallic Ni-Fe/SBA-15 catalysts have been synthesized and tested for the LA hydrogenation to GVL by the MPV route, in the presence of MeOH as the only hydrogen donor, and the oxophilic promotion of Fe was exploited to decrease the reduction temperature of the NiO phase and improve the reducibility of mixed oxides. Ni, Fe, and bimetallic Ni-Fe catalysts with different Fe/Ni molar ratios have been prepared by impregnation on the SBA-15 support, which helps the LA conversion to GVL. Moreover, this support provides high specific surface area (ranging between 690 and 1040 m2/g), narrow pore size distribution (average pore diameter between 5 and 30 nm), and high thermal stability (wall thicknesses between 3 and 5 nm) [28,29]. To investigate the relationship between catalyst structure and activity, characterization of the bimetallic Ni-Fe catalysts was performed by X-ray powder diffraction (XRD), transmission electron microscopy (TEM), hydrogen temperature-programmed desorption (H_2-TPD), hydrogen temperature-programmed reduction (H_2-TPR), X-ray photoelectron spectroscopy (XPS), and in situ Fourier transform–infrared spectroscopy (FT-IR). Lastly, the effects of the main reaction parameters on the GVL production have been evaluated and discussed, starting from the results of preliminary catalytic tests.

2. Results and Discussion

2.1. Characterization of Catalysts

First, the catalysts have been characterized by the XRD technique to investigate the atomic structure of the particles (Figure 1). The XRD patterns of Ni/SBA-15 catalyst exhibit three diffraction peaks at 2θ values of 44.2°, 51.8°, and 76.3°, which are assigned to (111), (200), and (220) reflections of metallic Ni, respectively. Fu et al. reported analogous diffraction peaks for the Ni/Al_2O_3 catalyst at 2θ angles of 44.3°, 51.7°, and 76.3°, assigned to metallic Ni [15]. Nevertheless, the absence of the diffraction peak at 37.3°, due to the NiO species, confirms the complete reduction of the Ni particles [15]. In the case of the monometallic Fe/SBA-15 catalyst, another diffraction peak was found at about 44.7°, which can be assigned to the (111) Fe cubic planes [30]. However, in the case of the Ni-Fe/SBA-15 catalysts, the Fe peak is shifted to lower 2θ, indicating a deep interaction between Ni and Fe particles to form face-centered cubic structures (fcc) of the Ni-Fe alloy, favored by the larger atomic radius of Fe, which increases the lattice constants [16].

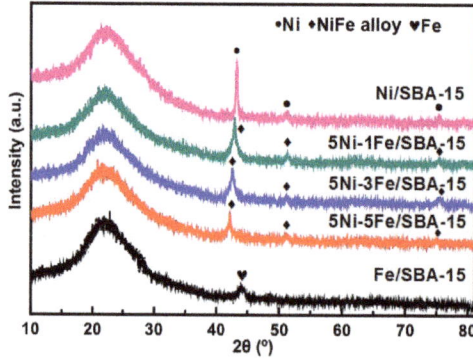

Figure 1. XRD (X-ray powder diffraction) patterns of the Ni/SBA-15, 5Ni-1Fe/SBA-15, 5Ni-3Fe/SBA-15, 5Ni-5Fe/SBA-15, and Fe/SBA-15 catalysts.

TEM characterization of the synthesized catalysts was carried out to better investigate their textural properties. As shown in Figure 2, Ni particles (black spots) are quite well dispersed on the porous SBA-15 support. For Ni/SBA-15, the average Ni particle size is about 20 nm, which is in agreement with data in the literature [31]. The addition of Fe decreases the average size of Ni particles, which are more uniformly distributed, especially in the case of 5Ni-3Fe/SBA-15 (Ni particle size of about 9 nm), indicating that the Fe doping promotes a more homogeneous Ni dispersion, minimizing the formation of aggregates. However, when both metals are used in the same amount (5:5), the average particle size becomes higher (16 nm), highlighting that the Ni:Fe molar ratio should be carefully tuned. Based on these preliminary characterization data, the optimum achieved Ni:Fe molar ratio is 5:3.

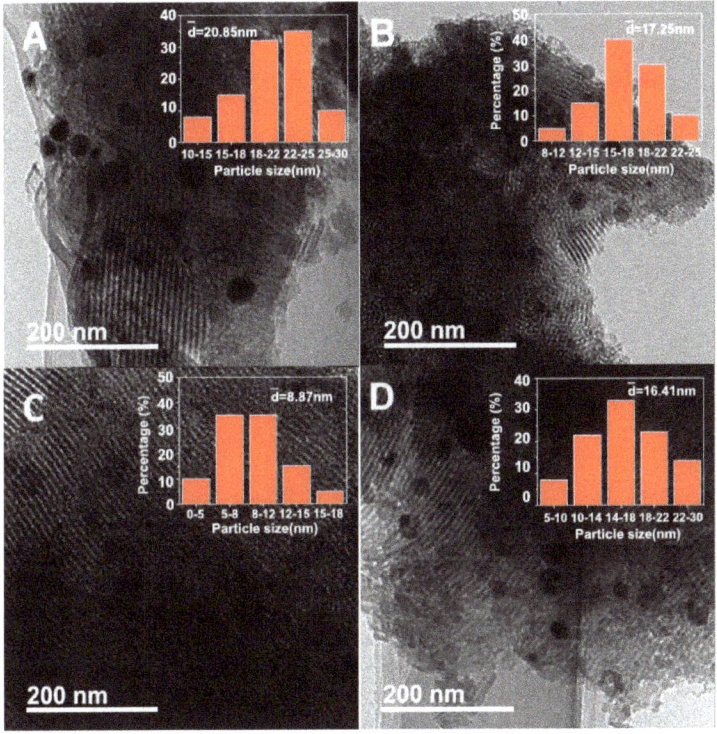

Figure 2. TEM (transmission electron microscopy) images of (**A**) Ni/SBA-15, (**B**) 5Ni-1Fe/SBA-15, (**C**) 5Ni-3Fe/SBA-15, and (**D**) 5Ni-5Fe/SBA-15 catalysts.

To further investigate the interactions between Ni and Fe within the Ni-Fe/SBA-15 catalyst, H_2-TPR analysis of both monometallic Ni/SBA-15, Fe/SBA-15, and bimetallic Ni-Fe/SBA-15 catalysts, was carried out (Figure 3). For the Ni/SBA-15 catalyst, only one peak was found at 430 °C, due to the reduction of NiO to metallic Ni. For the monometallic Fe/SBA-15 catalyst, three different reduction peaks were detected, at 370, 550, and 630 °C. In this regard, the reduction of Fe_2O_3 to α-Fe occurs by three consecutive steps: Fe_2O_3-Fe_3O_4-FeO-Fe [32]. Therefore, the peak at about 370 °C is due to the reduction from Fe_2O_3 to Fe_3O_4, whilst that from Fe_3O_4 to FeO occurs at about 550 °C, and that at 630 °C corresponds to the reduction from FeO to α-Fe [32]. Regarding the bimetallic Ni-Fe/SBA-15 catalyst, there are two different peaks between 300 and 550 °C, whilst no peak was found at 600 °C, thus confirming the absence of unalloyed Fe in the synthesized bimetallic catalyst. For 5Ni-1Fe/SBA-15, the reduction peak at 320 °C is assigned to the reduction of Fe_2O_3 to Fe_3O_4, because Fe_2O_3 is more easily reducible than NiO [33], whilst that at 440 °C is attributed to the reduction of NiO. Compared with the monometallic Ni/SBA-15 catalyst, the reduction peak of the Ni-Fe/SBA-15 catalysts is shifted to higher temperatures, corresponding to the increase of the Fe content, and the H_2 consumption gradually increases. H_2 may be more easily dissociated and activated on Ni, and then overflowed to the adjacent iron oxide to reduce it, thus leading to the formation of the Ni-Fe alloy [34].

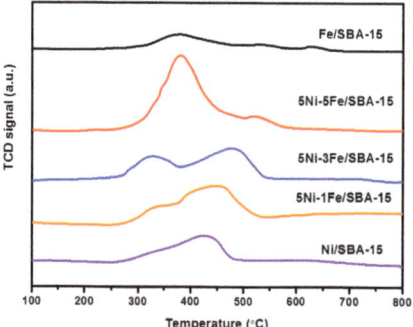

Figure 3. H_2-TPR (hydrogen temperature-programmed reduction) profiles of Ni/SBA-15, 5Ni-1Fe/SBA-15, 5Ni-3Fe/SBA-15, and 5Ni-5Fe/SBA-15 catalysts.

H_2-TPD profiles of the synthesized catalysts are reported in Figure S2. For Ni/SBA-15, a desorption peak was detected at 100 °C, in agreement with the adsorption of hydrogen on the surface of dispersed metal particles Ni in Figure S2. It is remarkable that the hydrogen desorption peak gradually moves to a higher temperature with the corresponding increase of Fe content. Moreover, the corresponding hydrogen desorption peaks for 5Ni-1Fe/SBA-15 and 5Ni-3Fe/SBA-15 are located at 105 and 125 °C, respectively, which is higher than that of Ni/SBA-15. However, 5Ni-5Fe/SBA-15 has two hydrogen adsorption sites, a low- (150 °C) and high-temperature desorption peak (300 °C), the latter deriving from the hydrogen desorption on the surface of Fe or from the poor dispersion of Ni. In addition, the hydrogen desorption peak from Ni/SBA-15 to 5Ni-5Fe/SBA-15 is shifted to a higher temperature, indicating that the adsorption of hydrogen on the active metal surface is enhanced. The desorption peak of Fe/SBA-15 appears at a higher temperature (250 °C), which is attributed to the hydrogen desorption from the Fe surface. The adsorption center corresponds to the desorption peak at low temperature, and desorption temperature of Fe/SBA-15 is the highest, meaning that the hydrogen desorption from the Fe species requires more energy than that from the Ni ones. To evaluate the dispersion of the particles, the amount of the hydrogen desorbed from the catalyst surface was calculated (Table 1), and it corresponds to the area of the desorption peak reported in Figure S2. The amount of desorbed hydrogen gradually increases with the area of the desorption peak, indicating the increase of the number of active hydrogen on the catalyst surface. With the increase in Fe content, that is moving from 5Ni-1Fe/SBA-15 to 5Ni-3Fe/SBA-15, the particle dispersion increases. However, for the 5Ni-5Fe/SBA-15 catalyst, the total amount of desorbed H_2 does not significantly increase, at the same time showing a worsening of the catalyst dispersion. Moreover, the total acidity of catalysts increases with the amount of Fe. It has been reported that, when Ni is doped with Fe, the unreduced Fe can act as a Lewis acid, thus increasing the acid content of the catalyst [35].

Table 1. Chemical properties of the synthesized monometallic Ni/SBA-15, Fe/SBA-15, and bimetallic Ni-Fe/SBA-15 catalysts.

Catalyst	Ni (mmol/g)	Fe (mmol/g)	H_2 (μmol/g) [a]	Particle Dispersion (%) [b]	Total Acidity (μmol/g) [c]
Ni/SBA-15	0.85	0	6.3	1.45	28.3
5Ni-1Fe/SBA-15	0.85	0.17	17.1	3.34	30.6
5Ni-3Fe/SBA-15	0.85	0.50	23.6	3.49	65.2
5Ni-5Fe/SBA-15	0.85	0.86	23.8	2.76	61.5
Fe/SBA-15	0	0.86	14.8	-	78.6

[a]: Amount of H_2 desorption calculated by TPD (Figure S2). [b]: H/(Ni + Fe) ratio. [c]: calculated by NH_3-TPD.

XPS analysis was performed to monitor the change of the oxidation state of the bimetallic catalysts. The XPS patterns of Ni 2p and Fe 2p of 5Ni/SBA-15, 5Ni-1Fe/SBA-15, 5Ni-3Fe/SBA-15, 5Ni-5Fe/SBA-15, and 5Fe/SBA-15 are shown in Figure S1, respectively. For the Ni 2p, Ni 2p $_{3/2}$ and 2p$_{1/2}$ spin orbit doublets were deconvoluted into six peaks within the range 845–885 eV (Figure S1a). For the 5Ni/SBA-15 catalyst, the peaks at 853.2, 856.7, and 861.7eV are assigned to the characteristic peaks of Ni^0, Ni^{2+}, and satellite peaks of Ni^{2+} in Ni 2P$_{3/2}$ orbit, respectively [8]. For 5Ni-1Fe/SBA-15, 5Ni-3Fe/SBA-15, and 5Ni-5Fe/SBA-15 catalysts, the characteristic peaks of Ni^0, Ni^{2+}, and satellite peaks of Ni^{2+} in Ni2P$_{3/2}$ orbit are progressively shifted to the higher binding energies, respectively, indicating a deeper interaction between Ni and Fe. In detail, the electrons are transferred from the oxophilic site of Fe to Ni, increasing the electron density of the outer layer of Ni, in agreement with the literature [36]. Moreover, the characteristic peaks of 5Ni-3Fe/SBA-15 are shifted greatly (0.6 eV), meaning that the reduction of 5Ni-3Fe/SBA-15 is certainly favored. For the Fe 2p in Figure S1b, two spin orbits are also present as Fe 2p$_{3/2}$ and Fe 2p$_{1/2}$. For the Fe/SBA-15 catalyst, the peaks at 707.7, 712.2, and 716.4eV are assigned to Fe^0, Fe^{2+}, and Fe^{3+} on Fe 2p$_{3/2}$, respectively [37]. For 5Ni-1Fe/SBA-15, 5Ni-3Fe/SBA-15, and 5Ni-5Fe/SBA-15 catalysts, the characteristic peaks of Fe^0 and Fe^{3+} are shifted towards the low binding energies, respectively, indicating that the electrons are indeed obtained by Fe, decreasing the electron density of the outer layer of Fe, and activating the C=O bond. From the XPS results, it can be concluded that the Lewis acid sites of the catalysts derive from the unsaturated coordination of Fe^{2+} and Fe^{3+} ions. Moreover, the characteristic peak area of Fe^0 in bimetallic Ni-Fe catalyst increases, indicating that the interaction between Ni and Fe promotes the reduction of iron oxide. The above results confirm that 5Ni-3Fe/SBA-15 is the most promising catalyst to test for the activation of the C=O bond of LA.

2.2. LA to GVL Catalytic Tests

Preliminary catalytic tests were carried out to test the effectiveness of the synthesized catalysts towards GVL production. First, GC-MS analysis was carried out to monitor the progress of the reaction (Figure S4). This analysis confirms that methyl levulinate and GVL represent the main intermediate and target product, respectively, indicating that LA could be completely hydrogenated to GVL, by properly tuning the reaction conditions. In this regard, 4-hydroxyvaleric acid, which is the intermediate of interest for the MPV intermolecular hydrogenation mechanism (path B, Scheme 2), was not detected in the reaction mixture. In principle, LA is firstly adsorbed on the surface of the catalyst and subsequently esterified with methanol (MeOH) to methyl levulinate, benefiting from the Lewis acidity of the catalyst [22]. To investigate the mechanism of LA adsorption on the catalysts, in situ FT-IR analysis was performed. As shown in Figure 4, the LA absorption band in the range 1750–1765 cm^{-1} is instead absent in Ni/SBA-15 and Ni-Fe/SBA-15, meaning that the physical adsorption of LA on these catalysts is weak. Moreover, in the case of the Ni/SBA-15 catalyst, an absorption band at about 1720 cm^{-1} was detected, due to the weak chemical adsorption of C=O stretching vibration of LA, indicating an interaction between the cationic or Lewis acid sites on the surface of Ni/SBA-15 catalyst and the lone pair electron of the carbonyl oxygen of LA [38]. In contrast, no absorption band was found at 1720 cm^{-1} for 5Ni-3Fe/SBA-15, whilst a blue shift occurs in the range 1650–1680 cm^{-1}, which is due to the strong LA chemisorption. The oxophilic promoter of Fe rendered more Lewis active sites for acid adsorption the C=O bond, in agreement with the results of XPS and NH$_3$-TPD characterization. Moreover, the composition of the gas phase was investigated to highlight possible differences between the catalysts towards the potential production of hydrogen (Figure S3). These analyses confirm that H$_2$ and CO$_2$ are the main reaction products of the gas phase. However, for the Ni/SBA-15 catalyst, the H$_2$ selectivity is only 40 mol %, whilst, in the case of 5Ni-3Fe/SBA-15, the addition of Fe significantly improves the H$_2$ selectivity up to 90 mol %. Instead, regarding the 5Ni-5Fe/SBA-15 catalyst, H$_2$ selectivity decreases, probably as a consequence of the larger particle size of Ni, according to the results of the above-discussed characterization.

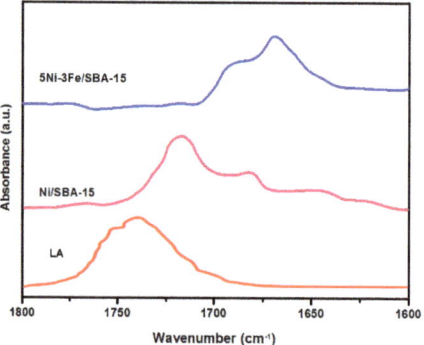

Figure 4. In situ FT-IR (Fourier transform–infrared spectroscopy) spectrum of LA and LA absorption on Ni/SBA-15 and 5Ni-3Fe/SBA-15 (50 mL·min^{-1} He, at 200 °C, 2 h).

The effect of temperature, reaction time, and hydrogen source on the LA hydrogenation to GVL was investigated. Figure 5a shows that the temperature has an important role on the LA conversion and GVL selectivity. Going from 50 to 100 °C, both LA conversion and GVL selectivity are low, whilst the catalytic performances significantly improve at 150 °C and, even better, at 200 °C. LA is not immediately hydrogenated into the corresponding alcohol, and methyl levulinate is the only detected intermediate, whilst GVL is subsequently produced by cyclization and dealcoholation mechanisms [39]. The effect of the reaction time was also considered (Figure 5b). The LA conversion is almost complete after only 30 min, and GVL selectivity rapidly increases from 30 to 120 min, becoming complete after 180 min. Lastly, the effect of the hydrogen source was preliminarily investigated by comparing the effect of molecular hydrogen and MeOH for the GVL production, in the absence/presence of the bimetallic 5Ni-3Fe/SBA-15 catalyst (Figure 5c). The LA conversion in MeOH is slightly higher than that obtained with molecular hydrogen. Remarkably, the GVL selectivity is significantly higher for the catalytic tests carried out in the presence of MeOH, thus confirming the positive synergism achieved thanks to the combined use of the bimetallic Fe-Ni catalyst and MeOH for promoting the selective hydrogenation of the C=O bond. These preliminary data confirm that the catalytic performances of the synthesized catalysts are comparable with those of noble metal catalysts, such as ruthenium, which has been identified as a very efficient catalyst for GVL production. For example, Xiao et al. reported an almost complete LA conversion (99.7 mol %) and a complete selectivity towards GVL, adopting a batch reactor, in the presence of Ru/graphene catalyst (40 bar H$_2$, 200 °C, 8 h) [40]. Piskun et al. found a maximum LA conversion of 90 mol %, but a low selectivity (66 mol % of GVL), using Ru/Beta-12.5 (45 Bar H$_2$, 90 °C, 2 h) [41]. Luo et al. reported a maximum GVL selectivity of 97.5 mol % at complete LA conversion, adopting Ru/TiO$_2$ as a catalyst, and dioxane as the reaction medium (4 MPa H$_2$, 200 °C, 4 h) [42]. However, in all these mentioned cases, pentanoic acid was identified as a reaction by-product, this compound deriving from the GVL ring opening, a reaction catalyzed by highly acidic supports [25].

Instead, in our case, pentanoic acid was not produced, thanks to the milder acidity of the adopted catalyst, leading to a more selective production of the desired GVL.

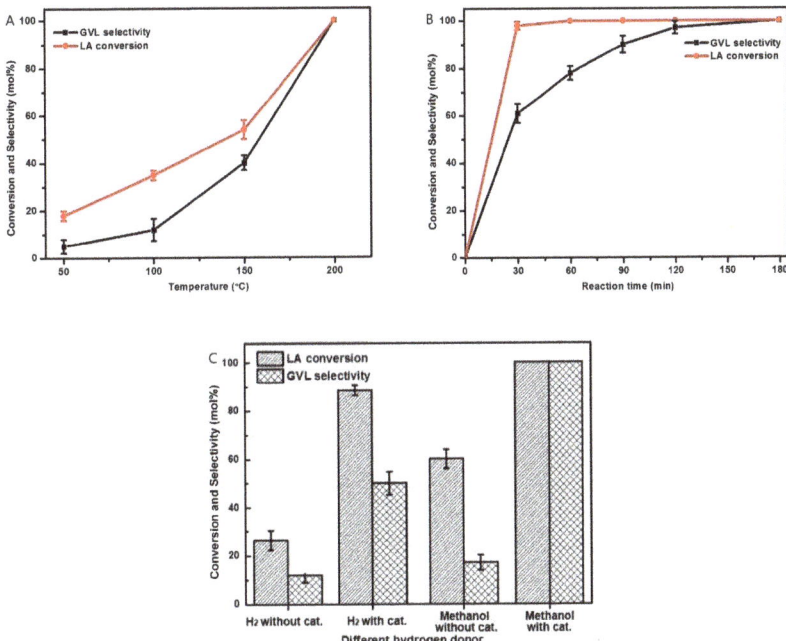

Figure 5. (**A**) The effect of temperature (reaction conditions: 2 g LA, 15 mL MeOH, 0.5 g 5Ni-3Fe/SBA-15, 180 min), (**B**) reaction time (reaction conditions: 2 g LA, 15 mL methanol, 0.5 g 5Ni-3Fe/SBA-15, 200 °C) and (**C**) different hydrogen sources (H_2 and MeOH) (reaction conditions: 2 g LA, 1 MPa H_2 or 15 mL MeOH, 0.5 g 5Ni-3Fe/SBA-15, 200 °C, 180 min) on LA conversion and GVL selectivity.

Preliminary recycling tests of the most promising bimetallic catalyst (5Ni-3Fe/SBA-15) were carried out to demonstrate its stability (Figure 6). The 5Ni-3Fe/SBA-15 catalyst was re-used consecutively five times under the best-identified reaction conditions. Before each cycle run, the spent catalyst was simply washed with ethanol in an ultrasonic cleaner and dried in a vacuum oven at 60 °C, thus avoiding tedious reactivation procedures. If compared with the performances of the fresh catalyst, those of the spent 5Ni-3Fe/SBA-15 catalyst decrease only slightly, even after five recycling tests, thus confirming the good potential of this catalytic system for GVL production on a larger scale.

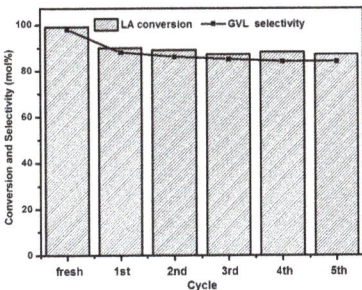

Figure 6. Recycling tests of 5Ni-3Fe/SBA-15 catalyst.

3. Experimental Section

3.1. Materials

Levulinic acid, γ-valerolactone, and methyl levulinate were purchased by Shanghai Aladdin Reagent Company (Aladdin, Shanghai, China). Nickel(II) nitrate hexahydrate [(Ni(NO$_3$)$_2$·6H$_2$O], ferric(III) nitrate hexahydrate [Fe(NO$_3$)$_3$·9H$_2$O], dichloromethane (DCM), and MeOH were purchased from Sigma-Aldrich Trading Co., Ltd. (Sigma-Aldrich, Shanghai, China). SBA-15 was produced by Sinopharm Chemical Reagent Co., Ltd. (Sinopharm, Shanghai, China). Double-distilled H$_2$O was used for the synthesis of catalysts.

3.2. Preparation of Catalyst

The metal-based catalysts were prepared by the incipient wetness impregnation method [21]. Solutions of nickel(II) nitrate hexahydrate and ferric(III) nitrate hexahydrate were mixed in the appropriate molar ratio (nominal weight content of 5 wt% Ni, Ni:Fe molar ratio = 5:0, 5:1, 5:2, 5:5). Then, 1 g of SBA-15 was added into the 10 mL volume of the metal precursor solution. The resulting slurry was stirred at room temperature for 2 h, then the catalyst was recovered, and dried overnight at 120 °C. The calcination was carried out under N$_2$ atmosphere at 450 °C (heating rate 5 °C/min) for 1 h and the catalyst was reduced at 450 °C (heating rate 5 °C/min) for 3 h, under H$_2$ flow.

3.3. Characterization of the Catalysts

Powder X-ray diffraction (XRD) analysis was carried out by a Rigaku Ultima IV X-ray diffractometer (Rigaku, Tokyo, Japan), utilizing Cu-Kα radiation (λ = 1.5405 Å). The test conditions were tube voltage, 50 kV; tube current, 50 mA; scanning speed, 8°/min; and scanning range, 10°–80°. N$_2$ sorption measurements were carried out at 77 K on a BEL-MAX gas/vapor adsorption instrument (MicrotracBEL Corp., Osaka, Japan). The surface areas were measured by the Brunauer-Emmett-Teller (BET) method. Transmission electron microscopy (TEM) was performed by the JEM-2100 microscope (JEOL, Tokyo, Japan), working at 200 kV. X-ray photoelectron spectroscopy (XPS) characterization was carried out by a Scientific Escalab 250 spectrometer (Thermo, Waltham, MA, USA) with AlKα (hν = 1486.6 eV) radiation, at a pressure of about 1 × 10^{-9} Torr. The binding energy values were corrected by using the C1s peak (284.6 eV). The total acidity was determined by NH$_3$-TPD technique. The calcined samples were characterized using temperature-programmed reduction by a TP-5080 adsorption instrument (xq-instrument, Tianjin, China), which was equipped with a TCD detector. Hydrogen temperature-programmed desorption (H$_2$-TPD) measurements were performed by an Autochem II-2920 (Norcross, GA, USA) instrument. The total H$_2$ desorption was calculated by measuring the areas of the desorption peaks. The FT-IR spectra were acquired by a Nicolet NEXUS 670 (Thermo, Waltham, MA, USA), which was equipped with an in situ IR cell.

3.4. Catalytic Tests

The catalytic hydrogenation of LA was carried out in a 100-mL stainless steel batch autoclave (Parr, Moline, IL, USA). In a typical run, the autoclave was loaded with the catalyst (0.5 g), LA (2 g), and MeOH (15 mL). Then, the reactor was heated up to the set-point temperature, under a constant stirring rate of 1000 rpm. At the end of the reaction, the slurry was recovered and centrifuged to separate the liquid phase from the spent catalyst, and the former was further analyzed by GC-MS and GC-FID techniques.

3.5. Analysis of the Liquid Phase

Qualitative analysis of the liquid samples was carried out by a GCMS-QP2010 instrument (Shimadzu, Kyoto, Japan), equipped with an RTX-5MS column (30 m × 0.25 mm × 0.25 μm). The injector temperature was 250 °C and a split ratio of 20:1 was employed. The MS operated in EI mode at

70 eV and the ion source temperature was set at 230 °C. The temperature program was the following: from 50 °C to 250 °C at a heating rate of 10 °C/min and held for 5 min at 250 °C. He was the carrier gas, and a column flow of 1.0 mL/min was used. The quantitative analysis was carried out by a GC-2010 gas-chromatograph (Shimadzu, Kyoto, Japan), equipped with an RTX-50 column (30 m × 0.25 mm × 0.25 μm) and a flame ionization detector. *n*-dodecane was used as the internal standard. Conversion and selectivity parameters were calculated according to the following equations:

$$\text{Conversion of LA (mol \%)} = \left(1 - \frac{n_{LA\ final}}{n_{LA\ initial}}\right) * 100\ \text{mol \%} \quad (1)$$

$$\text{Selectivity to GVL (mol \%)} = \frac{n_{GVL}}{n_{products}} * 100\ \text{mol \%} \quad (2)$$

4. Conclusions

γ-valerolactone is an important biomass platform molecule that can be used as a biofuel, pharmaceutical intermediate, food additive, and green solvent. The conversion of levulinic acid into high-value-added γ-valerolactone is an important reaction, and still requires the development of cheap and highly effective catalysts. In this regard, this paper is mainly focused on the synthesis of new Ni-Fe/SBA-15 bimetallic catalysts for improving the hydrogenation of levulinic acid into γ-valerolactone, adopting methanol as the only hydrogen donor. The physicochemical properties of the synthesized bimetallic catalysts were investigated by different techniques, e.g., XRD, TEM, XPS, and H_2-TPD. H_2-TPD confirmed that the addition of Fe as an oxophilic promoter lowers the reduction temperature of the NiO phase and improves the reducibility of the mixed oxides. At the same time, the presence of Fe favorably increased the Lewis acidity, which is beneficial for the MPV reduction of LA to GVL. Moreover, XPS analysis confirmed an electronic transfer from Fe to Ni, promoting the activation of the C=O bond of LA, and this conclusion is in agreement with the in situ FT-IR characterization. The preliminary catalytic tests have shown that the best bimetallic 5Ni-3Fe/SBA-15 catalyst has promising performances towards GVL synthesis. The effect of temperature and reaction time on GVL production was also considered, achieving complete LA conversion and GVL selectivity, working at 200 °C and after 180 min. Lastly, preliminary recycling tests of the best performing Ni-Fe bimetallic catalyst confirmed the possibility of its advantageous re-use, achieving satisfactory performances to γ-valerolactone, even after five catalytic cycles. The good catalytic performances/stability, the low cost, and the easy synthesis/regeneration of our Ni-Fe bimetallic catalysts are key aspects for the real development of γ-valerolactone production on a larger scale, thus filling the still existing gap between the industrial and academic world.

Supplementary Materials: The following are available online at http://www.mdpi.com/2073-4344/10/9/1096/s1. Figure S1: XPS profiles of reduced catalysts. Figure S2: H_2-TPD profiles of the synthesized SBA-15 supported catalysts. Figure S3: Influence of metals supported with SBA-15 on the amount of hydrogen in gas. Figure S4: GC-MS results for hydrogenation of LA with different reaction times.

Author Contributions: Conceptualization, L.L.; methodology, Q.Z. and X.H.; analysis, L.L., Q.Z. and X.H.; writing, original draft preparation, X.H. and Q.Z.; writing, review and editing, L.L.; project administration, L.L. All authors have read and agreed to the published version of the manuscript.

Funding: This work is supported by the National Natural Science Foundation of China (Grant No.31801321), and sponsored by Shanghai Sailing Program (Grant No. 19YF1436300) and Natural Science Foundation of Shanghai (Grant No. 18ZR1428100).

Conflicts of Interest: The authors declare no conflict of interest.

References

1. Climent, M.J.; Corma, A.; Iborra, S. Conversion of biomass platform molecules into fuel additives and liquid hydrocarbon fuels. *Green Chem.* **2014**, *16*, 516–547. [CrossRef]
2. Zhu, C.; Wang, H.L.Q.; Wang, C.; Xu, Y.; Zhang, Q.; Ma, L. Chapter 3-5-hydroxymethyl furfural-a C_6 precursor for fuels and chemicals. In *Biomass, Biofuels, Biochemicals*; Saravanamurugan, S., Pandey, A., Li, H., Riisager, A., Eds.; Elsevier: Amsterdam, The Netherlands, 2020; pp. 61–94. [CrossRef]
3. Kang, S.; Fu, J.; Zhang, G. From lignocellulosic biomass to levulinic acid: A review on acid-catalyzed hydrolysis. *Renew. Sustain. Energy Rev.* **2018**, *94*, 340–362. [CrossRef]
4. Molinari, V.; Antonietti, M.; Esposito, D. An integrated strategy for the conversion of cellulosic biomass into γ-valerolactone. *Catal. Sci. Technol.* **2014**, *4*, 3626–3630. [CrossRef]
5. Tang, X.; Sun, Y.; Zeng, X.; Lei, T.; Li, H.; Lin, L. γ-Valerolactone—An excellent solvent and a promising building block. In *Biomass, Biofuels, Biochemicals*; Saravanamurugan, S., Pandey, A., Li, H., Riisager, A., Eds.; Elsevier: Amsterdam, The Netherlands, 2020; pp. 199–226. [CrossRef]
6. Ye, L.; Han, Y.; Feng, J.; Lu, X. A review about GVL production from lignocellulose: Focusing on the full components utilization. *Ind. Crops Prop.* **2020**, *144*, 112031. [CrossRef]
7. Alonso, D.M.; Wettstein, S.G.; Dumesic, J.A. Gamma-valerolactone, a sustainable platform molecule derived from lignocellulosic biomass. *Green Chem.* **2013**, *15*, 584–595. [CrossRef]
8. Yan, Z.; Lin, L.; Liu, S. Synthesis of γ-valerolactone by hydrogenation of biomass-derived levulinic acid over Ru/C catalyst. *Energy Fuel* **2009**, *23*, 3853–3858. [CrossRef]
9. Manzer, L.E. Catalytic synthesis of α-methylene-γ-valerolactone: A biomass-derived acrylic monomer. *Appl. Catal. A Gen.* **2004**, *272*, 249–256. [CrossRef]
10. Son, P.A.; Nishimura, S.; Ebitani, K. Production of γ-valerolactone from biomass derived compounds using formic acid as a hydrogen source over supported metal catalysts in water solvent. *RSC Adv.* **2014**, *4*, 10525. [CrossRef]
11. Upare, P.P.; Lee, J.; Hwang, D.; Halligudi, S.B.; Hwang, Y.; Chang, J. Selective hydrogenation of levulinic acid to γ-valerolactone over carbon-supported noble metal catalysts. *J. Ind. Eng. Chem.* **2011**, *17*, 287–292. [CrossRef]
12. Deng, L.; Zhao, Y.; Li, J.; Fu, Y.; Liao, B.; Guo, Q. Conversion of levulinic acid and formic acid into γ-valerolactone over heterogeneous catalysts. *ChemSusChem* **2010**, *3*, 1172–1175. [CrossRef]
13. Yu, Z.; Lu, X.; Bai, H.; Xiong, J.; Feng, W.; Ji, N. Effects of solid acid supports on the bifunctional catalysis of levulinic acid to γ-valerolactone: Catalytic activity and stability. *Chem. Asian J.* **2020**, *15*, 1182–1201. [CrossRef] [PubMed]
14. Dutta, S.; Yu, I.K.M.; Tsang, D.C.W.; Ng, Y.H.; Ok, Y.S.; Sherwood, J.; Clark, J.H. Green synthesis of gamma-valerolactone (GVL) through hydrogenation of biomass-derived levulinic acid using non-noble metal catalysts: A critical review. *Chem. Eng. J.* **2019**, *372*, 992–1006. [CrossRef]
15. Mohan, V.; Venkateshwarlu, V.; Pramod, C.V.; Raju, B.D.; Rao, K.S.R. Vapour phase hydrocyclisation of levulinic acid to γ-valerolactone over supported Ni catalysts. *Catal. Sci. Technol.* **2014**, *4*, 1253–1259. [CrossRef]
16. Galletti, A.M.R.; Antonetti, C.; De Luise, V.; Martinelli, M. A sustainable process for the production of γ-valerolactone by hydrogenation of biomass-derived levulinic acid. *Green Chem.* **2012**, *14*, 688–694. [CrossRef]
17. Shimizu, K.; Kanno, S.; Kon, K. Hydrogenation of levulinic acid to γ-valerolactone by Ni and MoOx co-loaded carbon catalysts. *Green Chem.* **2014**, *16*, 3899–3903. [CrossRef]
18. Obregón, I.; Corro, E.; Izquierdo, U.; Requies, J.; Arias, P.L. Levulinic acid hydrogenolysis on Al_2O_3-based Ni-Cu bimetallic catalysts. *Chin. J. Catal.* **2014**, *35*, 656–662. [CrossRef]
19. Zhou, X.; Huang, Y.; Xing, W.; Liu, C.; Liao, J.; Lu, T. High-quality hydrogen from the catalyzed decomposition of formic acid by Pd- Au/C and Pd-Ag/C. *Chem. Commun.* **2008**, 3540–3542. [CrossRef] [PubMed]
20. Fellay, C.; Yan, N.; Dyson, P.J.; Laurenczy, G. Selective formic acid decomposition for high-pressure hydrogen generation: A mechanistic study. *Chem. Eur. J.* **2009**, *15*, 3752–3760. [CrossRef]
21. Ruppert, A.M.; Sneka-Płatek, O.; Jędrzejczyk, M.; Keller, N.; Dumon, A.S.; Michel, C.; Sautet, P.; Grams, J. Ru catalysts for levulinic acid hydrogenation with formic acid as a hydrogen source. *Green Chem.* **2016**, *18*, 2014–2028. [CrossRef]

22. Yu, Z.; Lu, X.; Bai, H.; Xiong, J.; Feng, W.; Ji, N. Heterogeneous catalytic hydrogenation of levulinic acid to γ-valerolactone with formic acid as internal hydrogen source: Key issues and their effects. *Chemsuschem* **2020**, *13*, 2916–2930. [CrossRef]
23. Chia, M.; Dumesic, J.A. Liquid-phase catalytic transfer hydrogenation and cyclization of levulinic acid and its esters to gamma-valerolactone over metal oxide catalysts. *Chem. Commun.* **2011**, *47*, 12233–12235. [CrossRef] [PubMed]
24. Gilkey, M.J.; Xu, B. Heterogeneous catalytic transfer hydrogenation as an effective pathway in biomass upgrading. *ACS Catal.* **2016**, *6*, 1420–1436. [CrossRef]
25. Luo, H.Y.; Consoli, D.F.; Gunther, W.R.; Román-Leshkov, Y. Investigation of the reaction kinetics of isolated Lewis acid sites in Beta zeolites for the Meerwein-Ponndorf-Verley reduction of methyl levulinate to γ-valerolactone. *J. Catal.* **2014**, *320*, 198–207. [CrossRef]
26. Tang, X.; Zeng, X.; Li, Z.; Li, W.; Jiang, Y.; Hu, L.; Liu, S.; Sun, Y.; Lin, L. In Situ generated catalyst system to convert biomass-derived levulinic acid to γ-Valerolactone. *Chemcatchem* **2015**, *7*, 1372–1379. [CrossRef]
27. Yi, Y.; Liu, H.; Xiao, L.; Wang, B.; Song, G. Highly efficient hydrogenation of levulinic acid into γ-valerolactone using an iron pincer complex. *Chemsuschem* **2018**, *9*, 1474–1478. [CrossRef] [PubMed]
28. Li, M.; Wei, J.; Yan, G.; Liu, H.; Tang, X.; Sun, Y.; Zeng, X.; Lei, T.; Lin, L. Cascade conversion of furfural to fuel bioadditive ethyl levulinate over bifunctional zirconium-based catalysts. *Renew. Energy* **2020**, *147*, 916–923. [CrossRef]
29. Norhasyimi, R.; Ahmad, Z.A.; Abdullah, R.M. A Review: Mesoporous santa barbara amorphous-15, types, synthesis and its applications towards biorefinery production. *Am. J. Appl. Sci.* **2010**, *7*, 1579–1586. [CrossRef]
30. Oyama, S.T.; Zhao, H.; Freund, H.J.; Asakura, K.; Włodarzyk, R.; Sierka, M. Unprecedented selectivity to the direct desulfurization (DDS) pathway in a highly active FeNi bimetallic phosphide catalyst. *J. Catal.* **2012**, *285*, 1–5. [CrossRef]
31. Jie, F.; Dong, S.; Lu, X. Hydrogenation of levulinic acid over nickel catalysts supported on aluminum oxide to prepare γ-valerolactone. *Catalysts* **2016**, *6*, 6. [CrossRef]
32. Hsu, P.; Jiang, J.; Lin, Y. Does a strong oxophilic promoter enhance direct deoxygenation? A study of NiFe, NiMo, and NiW catalysts in p-Cresol conversion. *ACS Sustain. Chem. Eng.* **2018**, *6*, 660–667. [CrossRef]
33. Ren, J.; Qin, X.; Yang, J.; Qin, Z.; Guo, H.; Lin, J.; Li, Z. Methanation of carbon dioxide over Ni-M/ZrO$_2$ (M=Fe, Co, Cu) catalysts: Effect of addition of a second metal. *Fuel Process. Technol.* **2015**, *137*, 204–211. [CrossRef]
34. Daniel, T.; José, L.P.; Isabel, S. Co-, Cu- and Fe-doped Ni/Al$_2$O$_3$ catalysts for the catalytic decomposition of methane into hydrogen and carbon nanofibers. *Catalysts* **2018**, *8*, 300. [CrossRef]
35. Nie, L.; Souza, P.M.D.; Noronha, F.B.; An, W.; Sooknoi, T.; Resasco, D.E. Selective conversion of m-cresol to toluene over bimetallic Ni-Fe catalysts. *J. Mol. Catal. A Chem.* **2014**, *388–389*, 47–55. [CrossRef]
36. Kumar, V.V.; Naresh, G.; Sudhakar, M.; Tardio, J.; Bhargava, S.K.; Venugopal, A. Role of brönsted and lewis acid sites on Ni/TiO$_2$ catalyst for vapour phase hydrogenation of levulinic acid: Kinetic and mechanistic study. *Appl. Catal. A Gen.* **2015**, *505*, 217–223. [CrossRef]
37. Kordulis, C.; Bourikas, K.; Gousi, M. Development of nickel based catalysts for the transformation of natural triglycerides and related compounds into green diesel: A critical review. *Appl. Catal. B Environ.* **2016**, *181*, 156–196. [CrossRef]
38. Ali, A.; Li, B.; Lu, Y.; Zhao, C. Highly selective and low-temperature hydrothermal conversion of natural oils to fatty alcohols. *Green Chem.* **2019**, *21*, 3059–3064. [CrossRef]
39. Nadgeri, J.M.; Hiyoshi, N.; Yamaguchi, A.; Sato, O.; Shirai, M. Liquid phase hydrogenation of methyl levulinate over the mixture of supported ruthenium catalyst and zeolite in water. *Appl. Catal. A Gen.* **2014**, *470*, 215–220. [CrossRef]
40. Xiao, C.; Goh, T.W.; Qil, Z.; Goes, S.; Brashler, K.; Perez, C.; Huang, W. Conversion of levulinic acid to γ-valerolactone over few-layer graphene-supported ruthenium catalysts. *ACS Catal.* **2016**, *6*, 593–599. [CrossRef]

41. Piskun, A.; Winkelman, J.G.M.; Tang, Z.; Heeres, H.J. Support screening studies on the hydrogenation of levulinic acid to γ-valerolactone in water using Ru catalysts. *Catalysts* **2016**, *6*, 131. [CrossRef]
42. Luo, W.; Deka, U.; Beale, A.M.; van Eck, E.R.H.; Bruijnincx, P.C.A.; Weckhuysen, B.M. Ruthenium-catalyzed hydrogenation of levulinic acid: Influence of the support and solvent on catalyst selectivity and stability. *J. Catal.* **2013**, *301*, 175–186. [CrossRef]

© 2020 by the authors. Licensee MDPI, Basel, Switzerland. This article is an open access article distributed under the terms and conditions of the Creative Commons Attribution (CC BY) license (http://creativecommons.org/licenses/by/4.0/).

Article

Guerbet Reactions for Biofuel Production from ABE Fermentation Using Bifunctional Ni-MgO-Al$_2$O$_3$ Catalysts

Zhiyi Wu, Pingzhou Wang, Jie Wang and Tianwei Tan *

Beijing Key Laboratory of Bioprocess, National Energy R&D Center for Biorefinery, College of Life Science and Technology, Beijing University of Chemical Technology, No. 15 of North Three-ring East Road, Chaoyang District, Beijing 100029, China; zywu@mail.buct.edu.cn (Z.W.); pzwang@mail.buct.edu.cn (P.W.); jiewang@mail.buct.edu.cn (J.W.)
* Correspondence: twtan@mail.buct.edu.cn

Abstract: To upgrade biomass-derived alcohol mixtures to biofuels under solvent-free conditions, MgO–Al$_2$O$_3$ mixed metal oxides (MMO) decorated with Ni nanoparticles (Ni–MgO–Al$_2$O$_3$) are synthesized and characterized. Based on the result, Ni nanoparticles are highly dispersed on the surface of MgAl MMO. As the Ni loading content varies from 2 to 10 wt.%, there is a slight increase in the mean Ni particle size from 6.7 to 8.5 nm. The effects of Ni loading amount, reducing temperature, and Mg/Al ratio on the conversion and product distribution are investigated. With the increase in both the Ni loading amount and reducing temperature, dehydrogenation (the first step of the entire reaction network) is accelerated. This results in an increase in the conversion process and a higher selectivity for the dialkylated compounds. Due to the higher strength and density of basic sites under high Mg/Al ratios, double alkylation is preferred and more long-chain hydrocarbons are obtained. A conversion of 89.2% coupled with a total yield of 79.9% for C$_5$–C$_{15}$ compounds is acquired by the as-prepared catalyst (prepared with Ni loading of 6 wt.%, reducing temperature of 700 °C, and Mg/Al molar ratio of 3. After four runs, the conversion drops by 17.1%, and this loss in the catalytic activity can be attributed to the decrease in the surface area of the catalyst and the increase in the Ni mean particle size.

Keywords: ABE fermentation; Ni-MgO-Al$_2$O$_3$ catalyst; biofuel; catalytic performance

Citation: Wu, Z.; Wang, P.; Wang, J.; Tan, T. Guerbet Reactions for Biofuel Production from ABE Fermentation Using Bifunctional Ni-MgO-Al$_2$O$_3$ Catalysts. *Catalysts* **2021**, *11*, 414. https://doi.org/10.3390/catal11040414

Academic Editor: Domenico Licursi

Received: 3 March 2021
Accepted: 17 March 2021
Published: 24 March 2021

Publisher's Note: MDPI stays neutral with regard to jurisdictional claims in published maps and institutional affiliations.

Copyright: © 2021 by the authors. Licensee MDPI, Basel, Switzerland. This article is an open access article distributed under the terms and conditions of the Creative Commons Attribution (CC BY) license (https://creativecommons.org/licenses/by/4.0/).

1. Introduction

The ever-growing concerns with regards to the continual depletion of fossil fuel reserves and the increasing severity of the environmental issues have urged the hastened development of clean energy sources. Biomass has received increasing attention as one of the most promising clean energy sources due to its abundant and renewable nature. In particular, the conversion of biomass into transportation fuels has emerged as a critical research focus in chemistry and engineering-related fields [1–12].

Triglyceride, starch-based, and lignocellulosic feedstocks are three types of feedstocks used in the production of biofuel [13–19]. Among these feedstocks, lignocellulosic biomass is considered the most promising candidate due to its high natural abundance [20–22]. To convert lignocellulose into biofuels, two steps are required. Step 1: Solid biomass has to be converted into platform chemicals with better catalytic activity to proceed with further treatments. Step 2: Conversion of platform chemicals into biofuels via C–C coupling reaction and hydrodeoxygenation (HDO), when necessary [23–25]. Various strategies can be implemented during the first step of the conversion process, whereby these strategies can be categorized into two parts: (1) Thermochemical process can be conducted under high temperature and/or pressure, and it can later be combined with chemical upgrading, e.g., Fischer–Tropsch synthesis. (2) The hydrolysis process can be carried out so that small molecules containing oxygen functional groups, such as acetone n-butanol-ethanol (ABE) fermentation products [26], can be obtained using biological or chemical means. After which, these generated small molecules can be upgraded into biofuels via catalytic reactions. Due to the ease of control on the molecular weight of the final hydrocarbons, the hydrolysis strategy has garnered more attention as compared with thermochemical strategy.

The chemical catalytic upgrading of ABE fermentation products to C_5–C_{11} ketones and alcohols in toluene was reported for the first time by Toste and coworkers [27]. The total yield was 86%. Then, the generated C_5–C_{11} ketones and alcohols were deoxygenated into components for the preparation of various products such as gasoline, jet fuel, and diesel fuel. Other than using toluene, Xu and coworkers [28] demonstrated the direct transformation of mimicking ABE fermentation products using water as the solvent, whereby Pd/C was coupled with various bases, e.g., K_3PO_4, KOH, and K_2CO_3, as the catalyst. In their report, the type and amount of bases used during the process can play pivotal roles in determining the total yield and product distribution. Furthermore, Xue et al. reported that concentrated ABE mixture could be directly alkylated to C_5–C_{15} or longer chain ketones in a continuous mode using a Pd/C catalyst, with an average conversion rate of >70% [29]. To further improve the recyclability and to avoid the use of alkaline as additives, metal supported on an alkaline substrate, e.g., hydrotalcite (HT) and CaO, has been developed [30,31]. For example, Lee and co-workers used Pd@C and CaO as solid bases to convert ABE mixture in a 180 °C batch reactor without any added solvent to produce a mixture of ketones and corresponding alcohols with 78% yield from acetone [32]. Among the various metallic materials (Ru, Pd, Fe, Co, Ni, Cu, and Zn) studied, Pd and Cu demonstrate the best performance with yields of 95% and 92%, respectively. On the other hand, Ni–HT catalyst shows a total yield of 2% [33]. Even though Pd-based and Cu-based catalysts demonstrate high performance, these catalysts still face challenges that require significant attention. For instance, besides the high cost of Pd, significant decarbonylation is observed for Pd–HT catalyst, which can lead to poor selectivity toward the desired products and carbon balance simultaneously [34]. As reported by Onyestyák and coworkers [35,36], a Cu-based catalyst is also unsuitable due to the high production of side products, i.e., esters, via Tishchenko reaction. As such, due to these limitations that plagued Pd-based and Cu-based catalysts, the development of cheap and efficient catalysts is urgently needed. In addition, to achieve green chemistry and simple separation, it is of great significance to convert ABE mixtures under solvent-free conditions. Furthermore, detailed investigations are greatly needed to provide insights into the reaction.

The reaction pathway is mainly comprised of dehydrogenation, aldol condensation, dehydration, and hydrogenation. This leads to clear design considerations when developing the catalysts: catalysts should possess (1) the ability to facilitate dehydrogenation of alcohols and (2) the capacity for aldol condensation. As reported in our previous work [37], Ni nanoparticles are regarded as the most promising catalyst for the upgrading of the ABE mixture due to their high catalytic activity for dehydrogenation/hydrogenation. It is well accepted that aldol condensation takes place at the acid–base site [38–41]. As a result, factors such as Ni loading, morphology of Ni nanoparticles, and the acidity–basicity of the catalyst can exert significant impacts on the catalytic activity of the catalyst.

Herein, a series of Ni–HT catalysts with different Ni loadings and acid–base properties is synthesized via the co-precipitation method. The as-prepared Ni–HT catalysts are characterized using scanning electron microscopy (SEM), high-resolution transmission electron microscope (HRTEM), X-ray diffraction (XRD), X-ray photoelectron spectroscopy (XPS), and temperature-programmed desorption (TPD). In this work, various parameters such as Ni loading content, temperature used in the reduction of catalyst, and Mg/Al ratio are systematically investigated for their effects on the total yield and product distribution.

2. Results and Discussion

2.1. Characterization of the as-Prepared Catalyst

XRD spectrums of Ni-MgO-Al_2O_3 catalysts with various Ni loading are presented in Figure 1. Distinct peaks located at 2θ = 36.1°, 43.1°, 62.6°, and 79.0° can be observed, which are consistent with those present in the standard XRD spectrum of MgO (JCPDS

01-075-1525). The diffraction peak located at 35.1° can be assigned to Al$_2$O$_3$, which overlaps with the diffraction peak of MgO. The characteristic peaks of Ni are not observed for 0 wt.% Ni loading, which is indicative that the pristine sample does not contain Ni. As the Ni loading increases to 2 wt.%, weak XRD peaks of Ni can be barely observed, which suggests that there is a low percentage of Ni in the catalyst with high dispersity. As the Ni loading increases, three distinct peaks at 44.5°, 51.8°, and 76.4° can be observed, which correspond to (111), (200), and (220) planes of metallic Ni (JCPDS 03-065-0380), respectively. This XRD result suggests the successful formation of Ni nanoparticles. As shown in Table 1, with the increase in Ni loading from 2 to 10 wt.%, the average crystallite size of Ni nanoparticle increases slightly from 6.7 to 8.5 nm, based on the Scherrer equation.

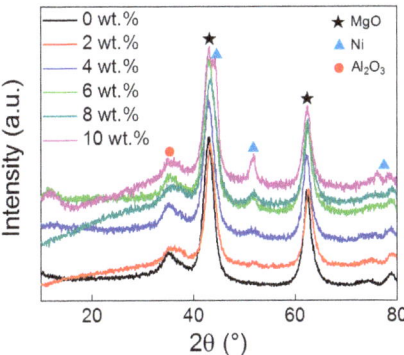

Figure 1. XRD spectrums of the as-prepared catalysts with various Ni loadings.

Table 1. Structural properties of Ni-MgO-Al$_2$O$_3$ catalysts with various Ni loadings.

Ni Loading (wt.%)	Surface Area (m^2/g)	Pore Volume (cm^3/g)	Mean Pore Diameter (nm)	Crystalline Size (nm)
0	267.1	0.8	6	/
2	256.1	0.74	5.8	6.7
4	238	0.69	5.7	7.1
6	237.5	0.68	5.7	7.5
8	237.4	0.65	5.5	7.9
10	227.2	0.63	5.5	8.5

To investigate the morphology of the as-prepared catalyst, SEM, and TEM are employed. Figure S1 shows the SEM images of the as-prepared catalyst. TEM images of various catalysts prepared with different Ni loadings of 2, 6, and 8 wt.% are shown in Figure 2a–c, respectively. Ni nanoparticles are clearly observed as dark spots in the TEM images, and they are highly dispersed on the surface of MgO-Al$_2$O$_3$. By measuring the size of more than 150 nanoparticles observed in the TEM images, corresponding histograms of Ni particle size distributions for catalysts with 2, 6, and 8 wt.% Ni loadings can be derived, and they are presented in Figure 3a–c, respectively. The average particle size of Ni nanoparticles is estimated based on a number-weighted diameter ($\bar{d} = \sum n_i d_i / \sum n_i$, n_i is the number of counted Ni particles with a diameter of d_i) with values of 6.8 and 7.8 nm for catalysts with 2 and 8 wt.% Ni loadings, respectively. This result confirms that the mean particle size of Ni nanoparticles increases slightly with the increase in Ni loading, which is consistent with the XRD results. Based on the high-resolution TEM image shown in Figure 2d, a lattice fringe of 0.203 nm can be clearly observed for the Ni nanoparticle, which corresponds to the (111) plane of Ni [42].

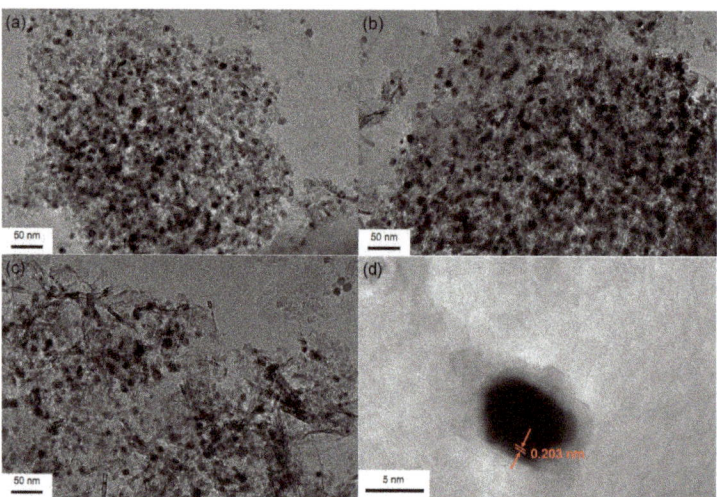

Figure 2. TEM images of Ni-MgO-Al$_2$O$_3$ catalysts with Ni loadings of (**a**) 2 wt.%, (**b**) 6 wt.%, and (**c**) 8 wt.% Ni; (**d**) HRTEM image of Ni-MgO-Al$_2$O$_3$ catalyst (6 wt.%).

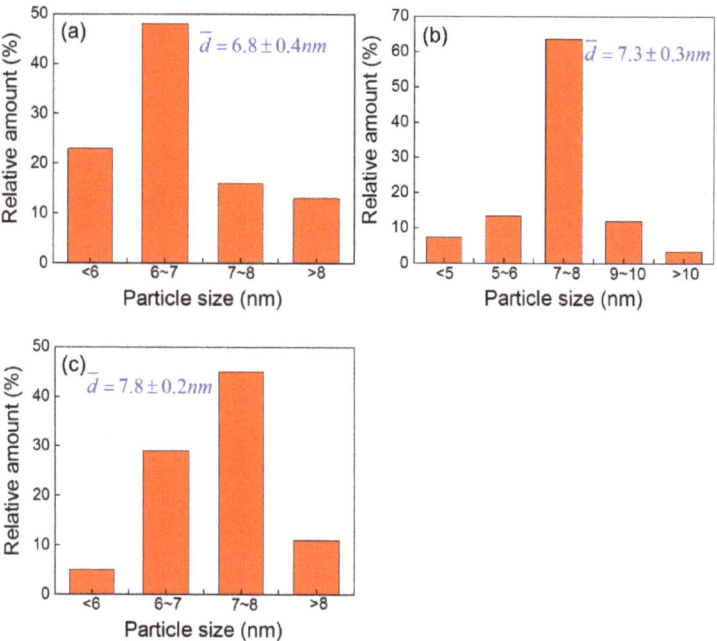

Figure 3. Histograms of the particle size distribution of Ni nanoparticles in the as-prepared catalysts with various Ni loadings: (**a**) 2 wt.%, (**b**) 6 wt.%, and (**c**) 8 wt.%.

To provide greater details to the dispersibility of Ni nanoparticles in the as-prepared catalyst, energy-dispersive X-ray spectroscopy (EDS) elemental mapping is conducted for Ni-MgO-Al$_2$O$_3$ catalyst with 6 wt.% Ni loading. As shown in Figure 4, the elemental distributions of Mg, Al, and Ni in the sample are highly uniform. This result suggests that Ni nanoparticles are homogeneously distributed across the well-mixed MgO-Al$_2$O$_3$ binary oxides.

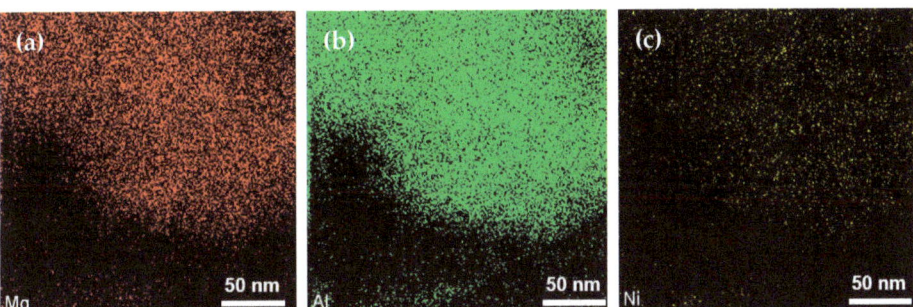

Figure 4. Energy-dispersive X-ray spectroscopy (EDS) elemental mapping of Ni-MgO-Al$_2$O$_3$ catalyst with 6 wt.% Ni loading. (**a**) Mg; (**b**) Al; (**c**) Ni.

BET surface areas, pore volumes, and pore size distributions of Ni-MgO-Al$_2$O$_3$ catalysts with various Ni loading are summarized in Table 1. It can be observed that all catalysts possess high surface areas (larger than 200 m^2/g), which is vital in providing a large contact area between the catalyst and the reactant, thus contributing toward high catalytic activity. With the increase in Ni loading, various parameters such as surface area, pore volume, and mean pore diameter exhibit a decreasing trend. Note that as Ni loading increases from 0 to 10 wt.%, the surface area of the sample decreases from 267.1 to 227.2 m^2/g. Meanwhile, the pore volume of the sample also decreases from 0.80 to 0.63 cm^3/g. This observation may be due to the increased occupancy of Ni nanoparticles in the sample as the Ni loading increases. Pore size distributions and N$_2$ isotherms are provided in Figure 5a,b, respectively. The pore size distribution, determined using the Barrett, Joyner, and Halenda method, illustrates that all the as-prepared catalysts contain mesopores with a mean pore size of approximately 6 nm. The N$_2$ adsorption–desorption isotherms of all the as-prepared catalysts reveal a typical Type IV isotherm with a well-defined N$_2$ hysteresis loop at relative pressures of 0.7–1.0. As such, based on this result, the as-prepared catalysts should possess mesoporous structures.

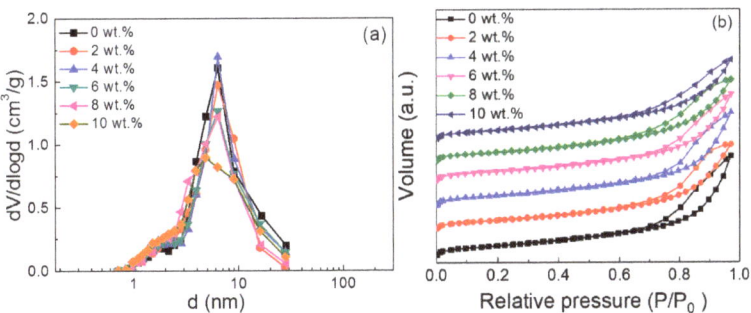

Figure 5. (**a**) Pore size distributions and (**b**) N$_2$ adsorption–desorption isotherms of Ni-MgO-Al$_2$O$_3$ catalysts with various Ni loadings.

2.2. Catalytic Upgrading of ABE Mixture

Scheme 1 shows the illustration of the catalytic upgrading mechanism of ABE mixtures to long-chain compounds. As illustrated in Scheme 1, three main reactions are involved during the catalytic upgrading process. Part A: Alkylation reactions producing ketones. Part B: Guerbet reactions generating alcohols with longer chains. Part C: Self-condensation of acetones through aldol condensation. Although the detailed mechanism of the Guerbet reaction is still controversial, it is generally believed that both the Guerbet reaction and alkylation reaction involve a series of processes, such as dehydrogenation of alcohol, aldol condensation between ketone (or aldehyde) and aldehyde, dehydration, and hydrogenation.

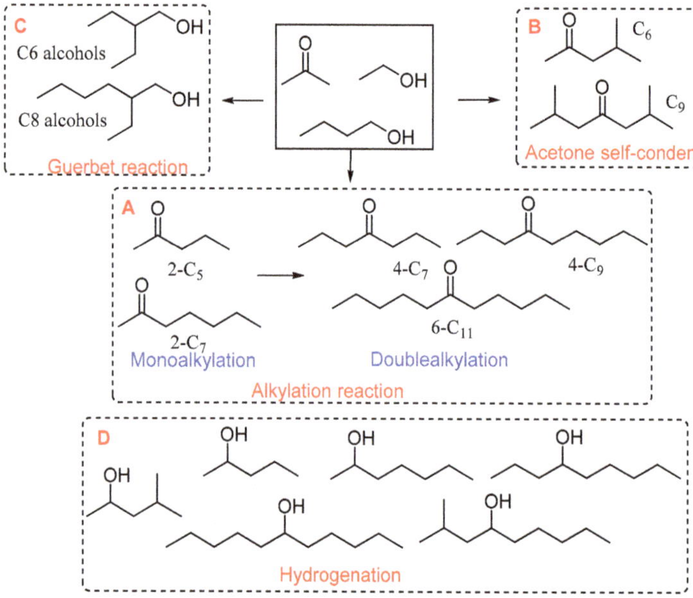

Scheme 1. Catalytic upgrading mechanism of ABE mixture.

It is commonly acknowledged that nanoparticles can facilitate the dehydrogenation/hydrogenation process. As such, loading amount and dispersibility of nanoparticles can play pivotal roles in influencing the catalytic activity of the catalyst. The growth of carbon chain molecules can be produced by aldol condensation. For example, aldol condensation between acetone and acetaldehyde yields 2-pentanone (2-C_5), and with further aldehyde condensation between 2-C_5 and acetaldehyde, 4-heptaneone (4-C_7) can be obtained. Part D: Alcohols (C5-OH, C7-OH, C9-OH, and C11-OH) can be generated via the hydrogenation of corresponding ketones as shown in Scheme 1. Ketone hydrogenation requires additional hydrogen resources, which may come from two ways. First, it may be released during the aldoesterification process (Scheme 2). However, neither ethyl acetate nor butyl butyrate is observed in our reaction system, and therefore these products can be ruled out. Second, steam reforming of ethanol or butanol may occur with the generation of hydrogen. As mentioned by Fu and Gong [43], nickel nanoparticles possess catalytic activity for the steam reforming of alcohols. The dehydrogenation of alcohols, decarbonylation of aldehydes, water–gas shift reaction, and CH_4 conversion are as follows. The activity of alcohol condensation reaction is highly related to the acidity and basicity of the catalyst, which indicates that the optimization of acid/base strength or acid/base amount of catalyst can play a significant role.

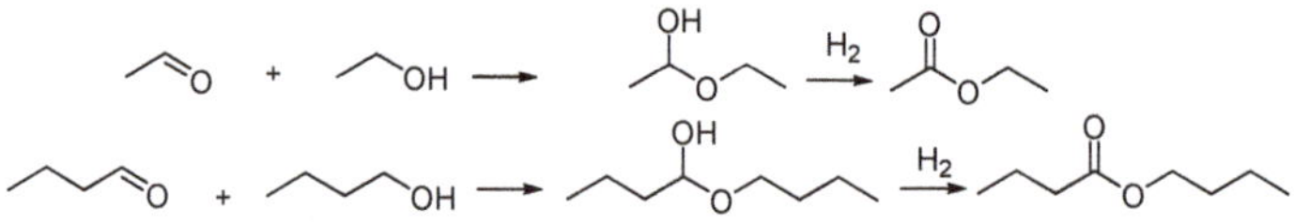

Scheme 2. Esterification between alcohol and aldehyde.

According to the abovementioned analyses, the effects of Ni amount on the ABE conversion yield and product distribution are firstly investigated. As shown in Figure 6, when ABE conversion is conducted using MMO catalyst (without Ni nanoparticle), an ABE conversion yield of 8.4% with two main products, i.e., 2-C_5 and 2-C_7 (monoalkylation of acetone with ethanol and butanol), are obtained. This result clearly shows that Mg-Al MMO exhibits low activity toward the dehydrogenation of ethanol and 1-butanol. Interestingly, as Ni nanoparticle is incorporated into Mg-Al MMO, ABE conversion yield is significantly improved as observed in Figure 6. Note that as Ni loading in the catalyst increases from 0 to 2 wt.%, ABE conversion yield increases drastically from 8.4% to 58.8%. After which, as Ni loading increases from 2 wt.% to 6 wt.%, ABE conversion yield continues to increase steadily from 58.8% to 89.2%. When the nickel loading is more than 6 wt.%, the yield decreases from 89.2% to 86.48% with increasing nickel loadings. Furthermore, it can be observed that with the increase in Ni loading, the main product changes from mono-alkylated compounds (C5 and 2-C7) to double-alkylated ones with longer carbon chains (4-C7 to C15). For instance, the total yield of double-alkylated compounds reaches 79.88% when the Ni loading is 6 wt.%. Two key reasons can be used to explain such phenomenon: (1) As Ni nanoparticle exhibits high dehydrogenation activity, increasing Ni loading would translate to the production of more aldehydes, which can then act as reactants for subsequent aldol condensation. This process can lead to a significant improvement in double alkylation. However, over-high Ni loading content is not valuable for the conversion yield. (2) C=C bonds in α, β-unsaturated ketones are kinetically and thermodynamically favored by the Ni site, and therefore saturated ketones are generated [44].

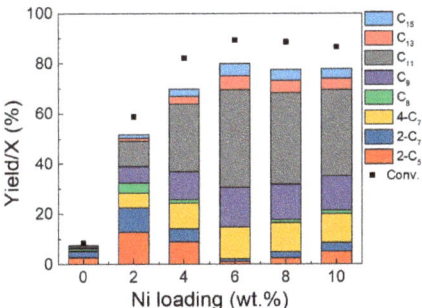

Figure 6. ABE conversion using Ni-MgO-Al_2O_3 catalysts with various Ni loading (reaction conditions: 15 g of ABE mixtures with 1.5 g of Ni-MgO-Al_2O_3 catalyst, molar ratio of Mg/Al: 3, temperature used in the reduction of catalyst: 700 °C, 240 °C, 20 h).

Other than the Ni loading contents, the temperature used in the reduction of the catalyst can also influence the catalytic activity of the catalyst. As such, various temperatures are used in the preparation process, and the corresponding catalytic performances of the as-prepared catalysts are shown in Figure 7. It is clearly shown that 7% ABE conversion yield with C5 as the sole product is achieved when catalysts that are reduced at 400 °C and 500 °C are used. As the temperature used in the reduction of catalyst increases to 600 °C, an increase in the ABE conversion yield is observed, with long-chain ketones and alcohols as the products. As the temperature increases from 600 °C to 800 °C, a total yield that increases from 68.4% to 88.6% is recorded. This observation is largely attributed to the fact that HT precursors would not be able to completely convert to MMO at a temperature lower than 500 °C, which results in higher catalytic activity for subsequent aldol condensation [45,46]. On the other hand, Ni nanoparticle is expected to catalyze the dehydrogenation of alcohols, while Ni^{2+} shows low dehydrogenation activity.

To verify the abovementioned hypothesis, XPS is used to further characterize the catalysts reduced at various temperatures. Binding energy values of metallic Ni are 852.7 eV (Ni $2p_{3/2}$) and 870.5 eV (Ni $2p_{1/2}$), while those of NiO are 854.0 eV (Ni $2p_{3/2}$) and 872.5 eV (Ni $2p_{1/2}$) [46,47]. As shown in Figure 8, as the temperature used in the reduction of the catalyst decreases to the range of 400 to 500 °C, a peak near 854.6 eV is observed, which indicates that Ni species exists primarily in the form of NiO. On the other hand, as the temperature increases from 600 to 800 °C, a peak around 852.6 eV can be clearly observed. The shift in the binding energy toward lower values may be attributed to the change in the configuration of Ni in the MgO-Al_2O_3 matrix. As indicated in Table 2, the amount of Ni nanoparticles in the catalyst increases from 7% to 87% as the temperature used in the reduction of catalyst increases from 400 to 800 °C.

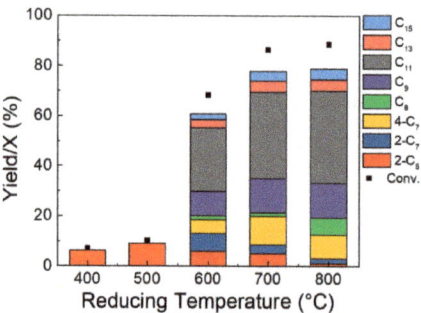

Figure 7. Effect of temperature used in the reduction of catalyst on its catalytic performance (reaction conditions: 15 g of ABE mixture with 1.5 g of Ni-MgO–Al_2O_3 catalyst, molar ratio of Mg/Al: 3, Ni loading amount: 6wt.%, 240 °C, 20 h).

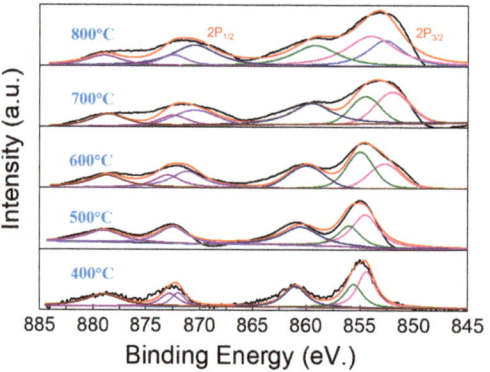

Figure 8. Ni 2p XPS spectrums of Ni-MgO–Al_2O_3 catalysts reduced at various temperatures.

Table 2. Quantity of oxidized and metallic Ni element in the catalysts reduced at various temperatures.

Temperature (°C)	Ni^0 Amount (%)	Ni^{2+} Amount (%)	Ni^0/Ni^{2+} Molar Ratio
400	7	93	0.07
500	17.4	82.6	0.21
600	55.7	44.3	1.28
700	78	22	3.55
800	87	13	6.69

It is well accepted that the acid–base properties of the catalyst can play a key role in influencing the aldol condensation activity. The weak Brønsted basic sites of MgAl-MMO are related to the residual surface hydroxyl groups after activation, the moderate strength Lewis sites are related to $Mg^{2-}O^{2-}$ and $Al^{3+}O^{2-}$ acid–base pairs, and the strong Lewis base sites are due to the existence of low coordinated O_2 species. The lower Al dopant content and the higher Mg content led to the increase in the basic center density, which is due to the formation of coordinated unsaturated oxygen sites. Materials with high Al contents are beneficial to the dehydration of alcohols rather than dehydrogenation and condensation. Ref. [48] by varying the Mg/Al-ratios, which changes the number and strength of the acid-base sites, the selectivity can be optimized towards dehydrogenation, aldolization, and hydride-shifts. Thus, the effect of the Mg/Al ratio on the catalytic performance of the catalyst is investigated. As shown in Figure 9, as the Mg/Al ratio increases from 1 to 9, ABE conversion yield remains constant at 88.4%. However, significant changes in the product distribution are observed across the varying Mg/Al ratio. Note that as the Mg/Al ratio increases from 1 to 9, the selectivity for C_5 and 2-C_7 decreases and more C_8–C_{15} are obtained. This result indicates that double alkylation is preferable at higher Mg/Al ratios. Figure S6 shows the results of the product distribution for the catalytic coupling of the ABE mixture.

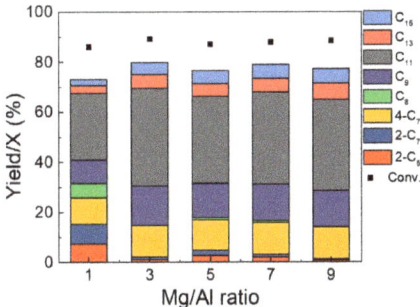

Figure 9. Effect of Mg/Al ratio on catalytic performance. Reaction conditions: 15 g of ABE mixture and 1.5 g of Ni-MgO-Al$_2$O$_3$ catalyst, Ni loading amount: 6 wt.%, temperature used in the reduction of catalyst: 700 °C, 240 °C, 20 h.

The acid and base properties of catalysts with various Mg/Al ratios are investigated using NH$_3$-TPD and CO$_2$-TPD, respectively. As shown in Figure 10a, catalysts with various Mg/Al ratios exhibit a similar profile with observable broad peaks at around 120 °C, which indicates that Ni-MgO-Al$_2$O$_3$ catalysts only contain weak acidic sites. As shown in Figure 10b, CO$_2$-TPD profiles are composed of two overlapping desorption peaks centered around 150 °C (peak I) and 260 °C (peak II), which correspond to weak and moderate basic sites, respectively.

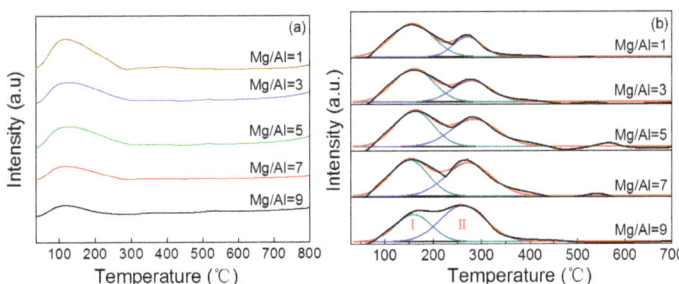

Figure 10. (**a**) NH$_3$-TPD and (**b**) CO$_2$-TPD profiles of Ni-MgO-Al$_2$O$_3$ catalysts with various Mg/Al ratios.

The concentrations of the acidic sites of the catalysts with various Mg/Al ratios are listed in Table 3. The concentration of acidic sites gradually decreases to a minimum value of 0.27 µmol/g with an increase in Mg/Al ratio. On the other hand, with the increase in Mg/Al ratio, the density of weak basic site decreases from 0.69 to 0.61 µmol/g, while densities of moderate basic site and the total basic site gradually increase. The weak acidic sites in the catalyst are beneficial toward the dehydration of unstable aldol products.

However, the presence of strong acid sites could potentially result in side reactions such as dehydration of alcohols to olefins. The basicity and number of basic sites play an important role in determining the product distribution. With the consideration of the results presented in Figure 9, it is suggested that there is a preferential double alkylation of acetone when using catalysts with higher basicity and more basic sites. This may explain the need for large amounts of alkali such as K_3PO_4, KOH, and K_2CO_3 in other works [27,28].

Table 3. Acid and base properties of Ni-MgO-Al_2O_3 catalysts with various Mg/Al ratios.

Mg/Al	Acid Sites (µmol/g)	Basic Sites (µmol/g) Total Basic Sites (mmol/g)		
		Weak Basic Sites (mmol/g)	Moderate Basic Sites (mmol/g)	Total Basic Sites
1:1	0.41	0.69	0.40	1.09
3:1	0.40	0.68	0.56	1.24
5:1	0.37	0.65	0.60	1.25
7:1	0.32	0.63	0.76	1.39
9:1	0.27	0.61	0.84	1.45

2.3. Regeneration Performance

The reusability of the as-prepared catalyst is studied, and the result is shown in Figure 11. It can be observed that ABE conversion yield decreases from 89.2% to 72.1% after four runs. To provide some insights into the decrease in the ABE conversion yield with runs, the spent catalysts are investigated with XRD, BET, and TEM. As shown in Figure S2, there is no distinct difference between the XRD spectrums of the fresh and spent catalysts, which indicates that the crystal structure and phase of the catalyst remain unchanged after the operation. The pore size distribution and N_2 adsorption–desorption isotherms of the spent catalysts are depicted in Figure S3, with the rest of the BET results listed in Table S1. With the increase in the number of recycling runs, the specific surface area and pore volume of the spent catalyst decrease gradually. Such a result may be the key factor towards the observed catalytic activity loss. TEM image of the spent catalyst is shown in Figure S4a, and the histogram of the size distribution of Ni nanoparticles is shown in Figure S4b. The mean diameter of Ni nanoparticles increases to 8.1 nm after several cycles, which is another factor that poses a detrimental effect on the catalytic performance of the catalyst. The base density of the used catalyst decreased significantly from 1.24 to 0 µmol/g, which may be caused by the formation of $MgCO_3$. The main peaks located at 400 °C and 540 °C can be attributed to the release of CO_2 from the decomposition of $MgCO_3$. The decrease in the base density and surface area of the activated catalyst may lead to a decrease in its catalytic activity.

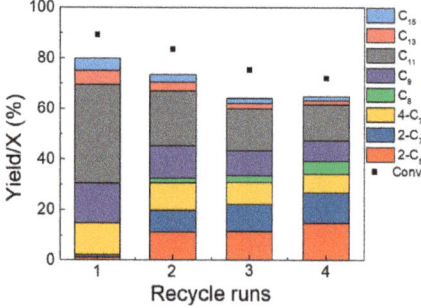

Figure 11. Regeneration performance of Ni-MgO-Al_2O_3 catalyst (reaction conditions: 15 g of ABE mixture and 1.5 g of Ni-MgO-Al_2O_3 catalyst, molar ratio of Mg/Al: 3, Ni loading: 6 wt.%, temperature used in the reduction of catalyst: 700 °C, 240 °C, 20 h).

3. Experimental Section

3.1. Materials

Al(NO$_3$)$_3$·9H$_2$O, Mg(NO$_3$)$_2$·6H$_2$O, and Ni(NO$_3$)$_2$·6H$_2$O were purchased from Sigma-Aldrich Co. (Sigma-Aldrich, St. Louis, MO, USA) Ethanol (99.9%), acetone (99.9%), and 1-butanol (99.9%) were purchased from Fuchen Chemical Plant (Tianjin, China). Deionized water was used in all reactions. All chemicals were used as received, without further purification.

3.2. Preparation of the Catalyst

Ni-MgO-Al$_2$O$_3$ mixed metal oxide (MMO) was prepared via co-precipitation. In a typical preparation process, an aqueous solution (0.1 L) of Na$_2$CO$_3$ (0.2 mol, 2.12 g) solution was added into an aqueous solution (0.1 L) containing Mg(NO$_3$)$_2$·6H$_2$O (1.05 mol, 26.92 g), Al(NO$_3$)$_3$·9H$_2$O (0.35 mol, 13.13 g), and Ni(NO$_3$)$_2$·6H$_2$O (0.09 mol, 2.62 g), and the mixture was mixed via vigorous stirring. The pH of the mixture was carefully maintained in the range of 9 to 10 using an aqueous 0.3 M NaOH solution. After continuously stirring and aging overnight, the precipitate (Mg-Al MMO) was filtered and washed with deionized water until the pH of the precipitate reached 7, and the washed precipitate was dried at 100 °C overnight. The as-prepared catalyst was subsequently dried at 80 °C overnight. Finally, the catalyst was reduced at 700 °C in the presence of H$_2$.

3.3. Characterization Techniques

XRD spectrum of the sample was recorded using a Shimadzu XRD-6000 diffractometer (Shimadzu, Tokyo, Japan) with Cu Kα radiation (λ = 1.5418 Å). A 2-theta value range of $5° \leq 2\theta \leq 90°$ was used in the XRD measurement. XPS spectrum of the sample was measured using a Thermo VGESCALAB 250 spectrometer (Thermo, Waltham, MA, USA), with Mg Kα (1253.6 eV) radiation as the X-ray source. All binding energies were calibrated with reference to the position of C1s peak at 284.6 eV. The morphology of the sample was observed under a ZEISS SUPRA55 SEM (ZEISS, Jena, Germany), with an accelerating voltage of 2.0 kV. The structure, size, and lattice fringes of the sample were examined under a JEOL JEM-2011 TEM (JEOL, Tokyo, Japan), equipped with an energy-dispersive X-ray spectrometer. The specific surface area of the sample was measured (Bruker, Karlsruhe, Germany) according to the Brunauer–Emmett–Teller (BET) method based on N$_2$ adsorption isotherm. All samples were degassed at 180 °C for 4 h prior to the BET measurement. The acid–base properties of the catalyst were determined using NH$_3$-TPD and CO$_2$-TPD, respectively, which are both equipped with a thermal conductivity detector (TCD, Bruker, Karlsruhe, Germany). 0.2 g sample was pretreated in a U-tube glass under a 25 mL/min He flow at 350 °C for 1 h, and it was then cooled to 100 °C. After completing the degassing of the sample, the adsorption gas was switched to CO$_2$ or NH$_3$, which was used to flush the U-tube glass for 30 min. After which, the temperature was increased from 100 to 850 °C at a heating rate of 10 °C/min, under a pure He atmosphere, to record the TCD signals.

3.4. Catalytic Conversion of Feedstock to Biofuel

A mixture of acetone, *n*-butanol, and ethanol with a molar ratio of 2.3:3.7:1 was used as a model of ABE fermentation. The catalytic conversion was conducted in a high-pressure reaction vessel, which was equipped with a magnetic stirring bar (IKA, Köln, Germany). In a typical reaction process, 1.5 g of the as-prepared catalyst and 15 g ABE mixture were added into the reactor, and the reactor was then heated to 240 °C for 20 h under a stirring speed of 800 rpm. After which, the reactor was cooled to room temperature by immersing it into an ice-water bath. The reaction product is analyzed using Shimadzu GC-2014 Chromatograph (Shimadzu, Tokyo, Japan) and Agilent GC-MS (Agilent, Palo Alto, Santa Clara, CA, USA) with DB-5 column, according to our previous work [37].

4. Conclusions

In summary, the potential of Ni-MgO-Al$_2$O$_3$ as a heterogeneous catalyst in biofuel production is investigated in this work. The as-prepared catalyst shows high efficiency in upgrading ABE mixture into long-chain (C$_5$–C$_{15}$) ketones and alcohols, which are important biofuel precursors. With the increase in Ni loading from 2 to 10 wt.%, the specific surface area of the catalyst decreases from 256.1 to 227.2 m^2/g, while the mean diameter of Ni nanoparticles increases from 6.7 to 8.5 nm. The acid–base properties of the as-prepared catalyst can be controlled by adjusting the Mg/Al molar ratio. Based on the result, catalysts with Mg/Al molar ratio in the range of 1 to 9 all show weak acidic sites, with a decreasing concentration of these acidic sites from 0.41 to 0.27 µmol/g. In contrast, as the Mg/Al molar ratio increases from 1 to 9 and the concentration of basic site increases from 1.09 to 1.45 µmol/g, with more moderate basic sites being generated. Furthermore, it is shown that higher conversion and greater preferential for double alkylation can be realized with higher Ni loading and higher temperature used in the reduction of catalyst. The Mg/Al molar ratio has little effect on the conversion yield, but it plays a significant role in influencing the product distribution. When employing a catalyst with a high amount of Mg, significant enhancement in the selectivity for 4-C$_7$ to C$_{15}$ hydrocarbons is observed. The catalyst, prepared with Mg/Al ratio of 3 and 6 wt.% Ni loading, and reduced at 700 °C, can achieve a conversion yield of 89.2% with the total C5–C15 compounds yield of 79.9%. The cyclic performance of the catalyst is also investigated, whereby there is a 17.1% decrease in the conversion yield after 4 runs. This loss in the activity may be a result of the decrease in the surface area and increase in the mean Ni particle size.

Supplementary Materials: The following are available online at https://www.mdpi.com/2073-4344/11/4/414/s1.

Author Contributions: Conceptualization, Z.W. and T.T.; writing—original draft preparation, Z.W., P.W. and J.W.; writing—review and editing, Z.W., P.W. and J.W.; funding acquisition, T.T. All authors have read and agreed to the published version of the manuscript.

Funding: The work described above was supported by the National Nature Science Foundation of China (21606008, U1663227), the Fundamental Research Funds for the Central Universities (ZY1630, JD1617, buctrc201616), and the State Key Laboratory of Chemical Engineering (SKL-ChE-17A02).

Data Availability Statement: The data presented in this paper are from published sources.

Acknowledgments: We gratefully acknowledge the support of the National Nature Science Foundation of China (21606008, U1663227), the Fundamental Research Funds for the Central Universities (ZY1630, JD1617, buctrc201616), and the State Key Laboratory of Chemical Engineering (SKL-ChE-17A02).

Conflicts of Interest: The authors declare no conflict of interest.

References

1. Alonso, D.M.; Bond, J.Q.; Dumesic, J.A. Catalytic Conversion of Biomass to Biofuels. *Green Chem.* **2010**, *12*, 1493–1513. [CrossRef]
2. Caes, B.R.; Teixeira, R.E.; Knapp, K.G.; Raines, R.T. Biomass to Furanics: Renewable Routes to Chemicals and Fuels. *ACS Sustain. Chem. Eng.* **2015**, *3*, 2591–2605. [CrossRef]
3. Liu, B.; Zhang, Z. Catalytic Conversion of Biomass into Chemicals and Fuels over Magnetic Catalysts. *ACS Catal.* **2016**, *6*, 326–338. [CrossRef]
4. Xia, Q.; Chen, Z.; Shao, Y.; Gong, X.; Wang, H.; Liu, X.; Parker, S.F.; Han, X.; Wang, Y. Direct hydrodeoxygenation of raw woody biomass into liquid alkanes. *Nat. Commun.* **2016**, *7*, 11162. [CrossRef]
5. Huber, G.W.; Iborra, S.; Corma, A. Synthesis of Transportation Fuels from Biomass: Chemistry, Catalysts, and Engineering. *Chem. Rev.* **2006**, *106*, 4044–4098. [CrossRef] [PubMed]
6. Corma, A.; Iborra, S.; Velty, A. Chemical Routes for the Transformation of Biomass into Chemicals. *Chem. Rev.* **2007**, *107*, 2411–2502. [CrossRef] [PubMed]
7. Demirbas, M.F. Biorefineries for biofuel upgrading: A critical review. *Appl. Energy* **2009**, *86*, S151–S161. [CrossRef]
8. Ghosh, S.; Chowdhury, R.; Bhattachary, P. Sustainability of cereal straws for the fermentative production of second generation biofuels: A review of the efficiency and economics of biochemical pretreatment processes. *Appl. Energy* **2017**, *198*, 284–298. [CrossRef]

9. Ennaert, T.; Aelst, J.V.; Dijkmans, J.; Clercq, R.D.; Schutyser, W.; Dusselier, M.; Verboekend, D.; Sels, B.F. Potential and challenges of zeolite chemistry in the catalytic conversion of biomass. *Chem. Soc. Rev.* **2016**, *45*, 584–611. [CrossRef]
10. Teng, J.J.; Ma, H.; Wang, F.R.; Wang, L.F.; Li, X.H. Catalytic Fractionation of Raw Biomass to Biochemicals and Organosolv Lignin in a Methyl Isobutyl Ketone/H_2O Biphasic System. *ACS Sustain. Chem. Eng.* **2016**, *4*, 2020–2026. [CrossRef]
11. Pileidis, F.D.; Titirici, M.M. Levulinic Acid Biorefineries: New Challenges for Efficient Utilization of Biomass. *ChemSusChem* **2016**, *9*, 562–582. [CrossRef]
12. Pang, J.F.; Zheng, M.Y.; Sun, R.Y.; Wang, A.Q.; Wang, X.D.; Zhang, T. Synthesis of ethylene glycol and terephthalic acid from biomass for producing PET. *Green Chem.* **2016**, *18*, 342–359. [CrossRef]
13. Zhang, Y.J.; Bi, P.Y.; Wang, J.C.; Jiang, P.W.; Wu, X.P.; Xue, H.; Liu, J.X.; Zhou, X.G.; Li, Q.X. Production of jet and diesel biofuels from renewable lignocellulosic biomass. *Appl. Energy* **2015**, *150*, 128–137. [CrossRef]
14. Han, J.; Sen, S.M.; Alonso, D.M.; Dumesic, J.A.; Maravelias, C.T. A strategy for the simultaneous catalytic conversion of hemicellulose and cellulose from lignocellulosic biomass to liquid transportation fuels. *Green Chem.* **2014**, *16*, 653–661. [CrossRef]
15. Liu, L.; Sun, J.S.; Cai, C.Y.; Wang, S.H.; Pei, H.S.; Zhang, J.S. Corn stover pretreatment by inorganic salts and its effects on hemicellulose and cellulose degradation. *Bioresour. Technol.* **2009**, *100*, 5865–5871. [CrossRef]
16. Liao, Y.H.; Liu, Q.Y.; Wang, T.J.; Long, J.X.; Ma, L.L.; Zhang, Q. Zirconium phosphate combined with Ru/C as a highly efficient catalyst for the direct transformation of cellulose to C6 alditols. *Green Chem.* **2014**, *16*, 3305–3312. [CrossRef]
17. Liao, Y.H.; Liu, Q.Y.; Wang, T.J.; Long, J.X.; Zhang, Q.; Zhang, Q.; Ma, L.L.; Zhang, Q.; Liu, Y.; Li, Y.P. Promoting Hydrolytic Hydrogenation of Cellulose to Sugar Alcohols by Mixed Ball Milling of Cellulose and Solid Acid Catalyst. *Energy Fuels* **2018**, *28*, 5778–5784. [CrossRef]
18. Dias, A.S.; Lima, S.; Pillinger, M.; Valente, A.A. Modified versions of sulfated zirconia as catalysts for the conversion of xylose to furfural. *Catal. Lett.* **2007**, *114*, 151–160. [CrossRef]
19. Mascal, M.; Dutta, S.; Gandarias, I. Hydrodeoxygenation of the Angelica Lactone Dimer, a Cellulose-Based Feedstock: Simple, High-Yield Synthesis of Branched C7–C10 Gasoline-like Hydrocarbons. *Angew. Chem. Int. Ed.* **2014**, *53*, 1854–1857. [CrossRef]
20. Wang, T.J.; Weng, Y.J.; Qiu, S.B.; Long, J.X.; Chen, L.G.; Li, K.; Liu, Q.Y.; Zhang, Q.; Ma, L.L. Gasoline Production by One-pot Catalytic Conversion of Lignocellulosic Biomass Derived Sugar/Polyol. *Energy Procedia* **2015**, *75*, 773–778. [CrossRef]
21. Liu, Y.; Chen, L.G.; Wang, T.J.; Zhang, Q.; Wang, C.G.; Yan, J.Y.; Ma, L.L. One-Pot Catalytic Conversion of Raw Lignocellulosic Biomass into Gasoline Alkanes and Chemicals over LiTaMoO6 and Ru/C in Aqueous Phosphoric Acid. *ACS Sustain. Chem. Eng.* **2015**, *3*, 1745–1755. [CrossRef]
22. Wang, J.J.; Liu, X.H.; Hu, B.C.; Lu, G.Z.; Wang, Y.Q. Efficient catalytic conversion of lignocellulosic biomass into renewable liquid biofuels via furan derivatives. *RSC Adv.* **2014**, *4*, 31101–31107. [CrossRef]
23. Yang, J.; Li, N.; Li, S.; Wang, W.; Li, L.; Wang, A.; Wang, X.; Yu, C.; Zhang, T. Synthesis of diesel and jet fuel range alkanes with furfural and ketones from lignocellulose under solvent free conditions. *Green Chem.* **2014**, *16*, 4879. [CrossRef]
24. Sheng, X.; Li, N.; Li, G.; Wang, W.; Wang, A.; Cong, Y.; Wang, X.; Zhang, T. Direct synthesis of gasoline and diesel range branched alkanes with acetone from lignocellulose. *Green Chem.* **2016**, *18*, 3707–3711. [CrossRef]
25. Xia, Q.N.; Cuan, Q.; Liu, X.H.; Gong, X.Q.; Lu, G.Z.; Wang, Y.Q. Pd/NbOPO4 Multifunctional Catalyst for the Direct Production of Liquid Alkanes from Aldol Adducts of Furans. *Angew. Chem. Int. Ed.* **2014**, *53*, 9755–9760. [CrossRef] [PubMed]
26. Cai, D.; Chang, Z.; Gao, L.L.; Chen, C.J.; Niu, Y.P.; Qin, P.Y.; Wang, Z.; Tan, T.W. Acetone–butanol–ethanol (ABE) fermentation integrated with simplified gas stripping using sweet sorghum bagasse as immobilized carrier. *Chem. Eng. J.* **2015**, *277*, 176–185. [CrossRef]
27. Anbarasan, P.; Baer, Z.C.; Sreekumar, S.; Gross, E.; Binder, J.B.; Blanch, H.W.; Toste, D. Integration of chemical catalysis with extractive fermentation to produce fuels. *Nature* **2012**, *491*, 235–239. [CrossRef]
28. Xu, G.Q.; Li, Q.; Feng, J.G.; Liu, Q.; Zhang, Z.J.; Wang, X.C.; Zhang, X.Y.; Mu, X.D. Direct a-Alkylation of Ketones with Alcohols in Water. *ChemSusChem* **2014**, *7*, 105–109. [CrossRef]
29. Xue, C.; Liu, M.; Guo, Y.; Hudson, E.P.; Chen, L.; Bai, F.; Liu, F.; Yang, S. Bridging chemica and bio-catalysis: High-value liquid transportation fuel production from renewable agricultural residues. *Green Chem.* **2017**, *19*, 660–669. [CrossRef]
30. Goulas, K.A.; Gunbas, G.; Dietrich, P.J.; Sreekumar, S.; Gippo, A.; Chen, J.P.; Gokhale, A.A.; Toste, F.D. ABE Condensation over Monometallic Catalysts: Catalyst Characterization and Kinetics. *ChemCatChem* **2017**, *9*, 677–688. [CrossRef]
31. Balakrishnan, M.; Sacia, E.R.; Sreekumar, S.; Gunbas, G.; Gokhale, A.A.; Scown, C.D.; Toste, F.D.; Bell, A.T. Novel pathways for fuels and lubricants from biomass optimized using life-cycle greenhouse gas assessment. *Proc. Natl. Acad. Sci. USA* **2015**, *112*, 7645–7649. [CrossRef]
32. Vo, H.T.; Yeo, S.M.; Dahnum, D.; Jae, J.; Hong, C.S.; Lee, H. Pd/C-CaO-catalyzed α-alkylation and hydrodeoxygenation of an acetone-butanol-ethanol mixture for biogasoline synthesis. *Chem. Eng. J.* **2017**, *313*, 1486–1493. [CrossRef]
33. Sreekumar, S.; Baer, Z.C.; Gross, E.; Padmanaban, S.; Goulas, K.; Gunbas, G.; Alayoglu, S.; Blanch, H.W.; Clark, D.S.; Toste, F.D. Chemocatalytic upgrading of tailored fermentation products toward biodiesel. *ChemSusChem* **2014**, *7*, 2445–2448. [CrossRef] [PubMed]
34. Goulas, K.A.; Sreekumar, S.; Song, Y.; Kharidehal, P.; Gunbas, G.; Dietrich, P.J.; Toste, F.D. Synergistic Effects in Bimetallic Palladium Copper Catalysts Improve Selectivity in Oxygenate Coupling Reactions. *J. Am. Chem. Soc.* **2016**, *138*, 6805–6812. [CrossRef] [PubMed]

35. Onyestyák, G.; Novodarszki, G.; Barthos, R.; Klebert, S.; Wellisch, A.F.; Pilbath, A. Acetone alkylation with ethanol over multifunctional catalysts by a borrowing hydrogen strategy. *RSC Adv.* **2015**, *5*, 99502–99509. [CrossRef]
36. Onyestyák, G.; Novodárszki, G.; Wellisch, F.; Pilbáth, A. Upgraded biofuel from alcohol-acetone feedstocks over a two-stage flow-through catalytic system. *Catal. Sci. Technol.* **2016**, *6*, 4516–4524. [CrossRef]
37. Zhu, Q.Q.; Shen, C.; Jie, W.; Tan, T.W. Upgrade of solvent-free acetone-butanol-ethanol mixture to high-value biofuels over Ni-containing Mgo-SiO$_2$ catalysts with greatly improved water-resistance. *ACS Sustain. Chem. Eng.* **2017**, *5*, 8181–8191. [CrossRef]
38. Carlini, C.; Marchionnab, M.; Noviello, M.; Gallettia, A.M.R.; Sbranaa, G.; Basiled, F.; Vaccarid, A. Guerbet condensation of methanol with n-propanol to isobutyl alcohol over heterogeneous bifunctional catalysts based on Mg-Al mixed oxides partially substituted by different metal components. *J. Mol. Catal. A Chem.* **2005**, *232*, 13–20. [CrossRef]
39. Horaa, L.; Kelbichováa, V.; Kikhtyanina, O.; Bortnovskiyb, O.; Kubicka, D. Aldol condensation of furfural and acetone over Mg Al layered double hydroxides and mixed oxides. *Catal. Today* **2014**, *223*, 138–147. [CrossRef]
40. Abello, S.; Medina, F.; Tichit, D.; Prez-Ramrez, J.; Groen, J.C.; Sueiras, J.E.; Salagre, P.; Cesteros, Y. Aldol Condensations Over Reconstructed Mg-Al Hydrotalcites: Structure-Activity Relationships Related to the Rehydration Method. *Chem. Eur. J.* **2005**, *11*, 728–739. [CrossRef]
41. Pupovac, K.; Palkovits, R. Cu/MgAl$_2$O$_4$ as Bifunctional Catalyst for AldolCondensation of 5-Hydroxymethylfurfural and SelectiveTransfer Hydrogenation. *ChemSusChem* **2013**, *6*, 2103–2110. [CrossRef]
42. Morgiela, J.; Szlezyngera, M.; Pomorskaa, M.; Marszałek, K.; Mani, R.; Marszałekb, K.; Mania, R. In-situ TEM heating of Ni/Al multilayers. *Int. J. Mater. Res.* **2015**, *106*, 703–710. [CrossRef]
43. Gong, Y.H.; Shen, C.; Wang, J.; Tan, W.T. Improved Selectivity of Long-Chain Products from Aqueous Acetone Butanol Ethanol Mixture over High Water Resistant Catalyst Based on Hydrophobic SBA-16. *ACS Sustain. Chem. Eng.* **2019**, *7*, 10323–10331. [CrossRef]
44. Pang, J.; Zheng, M.; He, L.; Li, L.; Pan, X.; Wang, A.; Wang, X.; Zhang, T. Upgrading ethanol to n-butanol over highly dispersed Ni–MgAlO catalysts. *J. Catal.* **2016**, *344*, 184–193. [CrossRef]
45. Carvalhoa, D.L.; Avillezb, R.R.; Rodriguesc, M.T.; Borgesa, L.; Appelc, L.G. Mg and Al mixed oxides and the synthesis of n-butanol from ethanol. *Appl. Catal. A Gen.* **2012**, *415–416*, 96–100. [CrossRef]
46. Xia, K.; Lang, W.; Li, P.; Yan, X.; Guo, Y. The properties and catalytic performance of Pt-In/Mg(Al)O catalysts for the propane dehydrogenation reaction: Effects of pH value in preparing Mg(Al)O supports by the co-precipitation method. *J. Catal.* **2016**, *338*, 104–114. [CrossRef]
47. Moulder, J.F.; Stickle, W.F.; Sobol, P.E.; Bomben, K.D. *Handbook of X-Ray Photoelectron Spectroscopy*; Physical Electronics: Chanhassan, MN, USA, 1995. [CrossRef]
48. Dries, G.; Willinton, Y.H.; Bert, S.; Pascal, V.D.V. Review of catalytic systems and thermodynamics for the Guerbet condensation reaction and challenges for biomass valorization. *Catal. Sci. Technol.* **2015**, *5*, 3876–3902. [CrossRef]

Article

A Novel and Efficient Method for the Synthesis of Methyl (*R*)-10-Hydroxystearate and FAMEs from Sewage Scum

Luigi di Bitonto, Valeria D'Ambrosio and Carlo Pastore *

Water Research Institute (IRSA), National Research Council (CNR), via F. de Blasio 5, 70132 Bari, Italy; luigi.dibitonto@ba.irsa.cnr.it (L.d.B.); valeria.dambrosio@ba.irsa.cnr.it (V.D.)
* Correspondence: carlo.pastore@ba.irsa.cnr.it

Abstract: In this work, the transesterification of methyl estolides (ME) extracted from the lipid component present in the sewage scum was investigated. Methyl 10-(*R*)-hydroxystearate (Me-10-HSA) and Fatty Acid Methyl Esters (FAMEs) were obtained in a single step. A three-level and four factorial Box–Behnken experimental design were used to study the effects of methanol amounts, catalyst, temperature, and reaction time on the transesterification reaction using aluminum chloride hexahydrate ($AlCl_3 \cdot 6H_2O$) or hydrochloric acid (HCl) as catalysts. $AlCl_3 \cdot 6H_2O$ was found quite active as well as conventional homogeneous acid catalysts as HCl. In both cases, a complete conversion of ME into Me-10-HSA and FAMEs was observed. The products were isolated, quantified, and fully characterized. At the end of the process, Me-10-HSA (32.3%wt) was purified through a chromatographic separation and analyzed by NMR. The high enantiomeric excess (ee > 92%) of the *R*-enantiomer isomer opens a new scenario for the valorization of sewage scum.

Keywords: sewage scum; methyl (*R*)-10-hydroxystearate; FAMEs; biodiesel; estolides

Citation: di Bitonto, L.; D'Ambrosio, V.; Pastore, C. A Novel and Efficient Method for the Synthesis of Methyl (*R*)-10-Hydroxystearate and FAMEs from Sewage Scum. *Catalysts* **2021**, *11*, 663. https://doi.org/10.3390/catal11060663

Academic Editor: Domenico Licursi

Received: 24 April 2021
Accepted: 21 May 2021
Published: 23 May 2021

Publisher's Note: MDPI stays neutral with regard to jurisdictional claims in published maps and institutional affiliations.

Copyright: © 2021 by the authors. Licensee MDPI, Basel, Switzerland. This article is an open access article distributed under the terms and conditions of the Creative Commons Attribution (CC BY) license (https://creativecommons.org/licenses/by/4.0/).

1. Introduction

Hydroxy Fatty Acids (HFAs) are valuable raw materials widely used for several industrial applications, including resins, polymers, cosmetics, biofuels, biolubricants, and additives in coatings and paintings [1,2]. They are valuable intermediates for synthesizing chemicals and pharmaceuticals for their antibiotic, anti-inflammatory, and anticancer properties [3,4].

HFAs are ubiquitous as constituents of plants, seeds, insects, animals and other microorganisms [5,6]. Many of these natural sources are found as part of estolides, oligomeric fatty acid esters formed by hydroxy acyl groups bonded together with ester bonds [7]. Estolides are being marketed as biolubricants for automotive and industrial applications for their excellent physicochemical properties as high viscosity and flash point, good resistance and biodegradability [8–10].

Since its first discovery, 10-(*R*)-Hydroxystearic acid (10-HSA) has attracted great industrial interest. It is the natural precursor of γ-(*R*)-dodecalactone, a taste and aroma component used in the flavor and fragrance industry [11–13]. Moreover, it is used in the manufacturing of lubricants and cosmetics for its chemical properties similar to those of Ricinoleic acid (or 12-Hydroxystearic acid) [14,15].

In recent years, different studies have been carried out for the production of 10-HSA, based on the enzymatic hydrolysis of vegetable oils from bacteria and microorganisms, such as *Elizabethkingia meningoseptica* [16], *Enterococcus faecalis* [17], *Lactobacillus plantarum* [18], *Lysinibacillus fusiformis* [19], *Nocardia cholesterolicum* [20], *Selenomonas ruminantium* [21], *Stenotrophomonas nitritireducens* [22], *Stenotrophomonas maltophilia* [23], *Sphingobacterium thalpophilum* [24]. Fatty acid hydratases have shown to be efficient catalysts with a good regio- and stereoselectivity, particularly useful to obtaining pure enantiomeric forms [25,26]. However, their applicability is not competitive with the currently existing conventional diesel-producing technology for a series of drawbacks, including (i) the specificity of the substrate, which can be a problem for the conversion of some feedstocks, (ii) the instability of the enzymes in the organic solvents, often required for the solubilization of reacting substrate or recovery of the final product, and (iii) their excessive cost [27].

The use of edible oils for human consumption represents a non-sustainable choice from an economic and environmental perspective. The development of new solutions to produce 10-HSA or its derivatives, mainly if based on non-edible feedstocks, could be a challenging goal for economic reasons and environmental and ethical concerns. Although considered a waste, sewage scum can be used as a source of energy and resources, thus replacing non-renewable resources with a considerable environmental impact [28–31].

Identified as CER190809, it is a floatable by-product of the wastewater treatment plants (WWTPs) obtained from the primary and secondary settler tanks. It mainly consists of vegetable oils and grease, animal fats, and food waste deriving from households, restaurants and animal product industries [32]. Due to their low density, these oily materials float on the wastewater surface. They can be easily skimmed off at the beginning of the treatment processes and used for energy purposes. Usually, sewage scum is processed in the anaerobic digester to produce biogas for electricity generation in the same plant [33]. However, the separation of sewage scum is often avoided compared with that of primary and secondary sludge, which instead accounts for about 50% of the total operating costs of a WWTP [34,35]. More frequently, they are directly disposed of in landfills, increasing the cost of treatment facilities and negatively impacting the environment. For these reasons, researchers have focused on developing novel technologies for their full exploitation. The use of sewage sludge as a lipid feedstock for biodiesel production is an alternative and sustainable approach to sludge management and disposal challenges [36].

Biodiesel is a biodegradable and renewable fuel with chemical and physical properties similar to petroleum-based fuels [37–39]. It is a mixture of Fatty Acid Methyl Esters (FAMEs), which can be synthesized by the reaction of different lipid fractions with methanol in the presence of an acid, a base or an enzyme catalyst [40,41]. The main obstacle to its marketing is the raw materials (mostly vegetable oils and animal fats), which constitute about 70–85% of production costs [42,43]. As a result, the use of non-edible alternative oils is constantly growing. Lipids extracted from sewage scum are mainly constituted by Free Fatty Acids (FFAs, 45–55%wt) and calcium soaps of fatty acids (25–30%wt) [28,29]. After the chemical activation with formic acid [44], they can be easily converted into the corresponding methyl esters by direct esterification with methanol using aluminum chloride hexahydrate (AlCl$_3$·6H$_2$O) as a catalyst [45]. However, besides the biodiesel production (75–80%wt), methyl estolides (ME, 15–20%) were also isolated and characterized [28]. These last, obtained as a result of bacteria activity in sewage sludge [46], can be further converted into methyl-10-hydroxystearate (Me-10-HSA) and FAMEs, representing a valuable source for 10-HSA production (Figure 1) through a transesterification reaction.

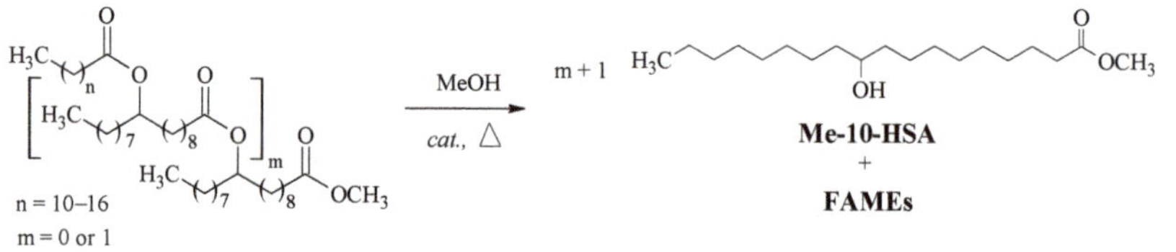

Figure 1. Schematic process of the transesterification reaction of methyl estolides with methanol for the synthesis of Me-10-HSA and FAMEs.

Such a process would contribute to the production of Me-10-HSA from non-edible feedstocks and a complete valorization of the lipid fraction present in the sewage scum. In this study, an acidic transesterification was proposed for the synthesis of Me-10-HSA and FAMEs by direct conversion of ME isolated from sewage scum with methanol. Aluminum chloride hexahydrate (AlCl$_3$·6H$_2$O) and hydrochloric acid (HCl) were used as catalysts. The best operative conditions were determined through a response surface methodology, widely adopted in studies concerning the production of biodiesel for optimizing the transesterification reaction [47–50]. Notably, a Box–Behnken factorial design of experiments was used. The amount of methanol and catalyst, temperature and reaction time were optimized with the aim of maximizing the conversion of ME into Me-10-HSA and FAMEs.

2. Analysis of Results

2.1. Characterization of the Lipid Component of Sewage Scum and Biodiesel Production

Biofuels are considered the leading renewable energy sources, presenting several advantages with respect to conventional fossil fuels [51,52]. Nevertheless, the high production costs associated with the raw materials (vegetable oils and animal fats) result in a significant increase in their price [42,43].

Sewage scum can be considered a cheap and available feedstock to synthesize biofuels due to its high lipid content (up to 36–50% of dry weight) [29]. The lipid fraction, very rich in FFAs, can be easily converted into FAMEs by acid-catalyzed direct esterification [53–55]. However, the high water content in the sewage scum (TS = 10–25%wt) represents a significant obstacle to biodiesel production at a commercial scale. The initial stages, from the collection of the raw sludge to the dehydration and drying, are expensive processes, which make the biodiesel production from sewage scum not economically feasible [56]. Furthermore, the subsequent extraction of the lipid fraction requires a significant amount of organic solvent, thus increasing the manufacturing costs [57].

Lastly, the method typically known for biodiesel production from sewage scum is based on homogeneous acid catalysts as H_2SO_4 [53–55]. Still, it is not competitive with the conventional technologies from triglycerides under alkaline catalysis for a series of drawbacks: (i) the recovery of the catalyst takes place only partially, and (ii) additional steps are required for the purification of the final products. A new methodology was then developed to successfully convert wastewater sewage scum into biodiesel, consisting of four different steps [28]. The overall process is outlined in Figure 2. First, sewage scum was heated at a temperature of 80 °C, with the lipid component was recovered by centrifugation at 4000 rpm for 3 min (recoverability > 90%), without the addition of solvents or acids [29]. Subsequently, the lipid extract was activated by adding the stoichiometric amount of formic acid (HCOOH) to calcium soaps (25–30%), thus obtaining their complete conversion into FFAs [44]. Activated lipids (FFAs = 75–80%) were then efficiently converted into the corresponding methyl esters by direct esterification using $AlCl_3 \cdot 6H_2O$ as a catalyst [45].

As a result, about 95% FFAs were converted into FAMEs with minimal reactants under mild conditions (molar ratio FFAs:MeOH:catalyst = 1:10:0.02, 72 °C, 2 h). Moreover, the use of $AlCl_3 \cdot 6H_2O$ favored a convenient separation of products between the two phases: the catalyst was recovered entirely, with the upper methanol phase along with the water produced during the reaction, whereas the methyl esters were present in the lower oily phase [45]. This resultant oily phase was recovered and pure FAMEs (75–80%wt) were collected by vacuum distillation (Figure 2).

The proposed scheme for exploiting sewage scum for biodiesel production has proven to be economically viable and applicable on an industrial scale. Nevertheless, the potential of the lipid component has not yet been fully exploited. After the distillation process, a residue was recovered (20–25%wt), which was analyzed by preparative chromatography. The residue was mainly composed of: ME (50.3%wt), polar compounds (33.8%), FAMEs (6.4%wt) and small quantities of Me-10-HSA (3.6%wt), Mineral oils (2.7%wt), Waxes (1.8%wt), FFAs (1.0%wt) and Methyl-10-ketostearate (0.4%wt). ME already have a potential market value as biolubricants [8–10]; however, to obtain a complete valorization of the lipid component, a further improvement of the reaction by-products could help to improve the economy of the overall process. For these reasons, the transesterification reaction of ME with methanol for the synthesis of Me-10-HSA and FAMEs was investigated, by optimizing the process parameters.

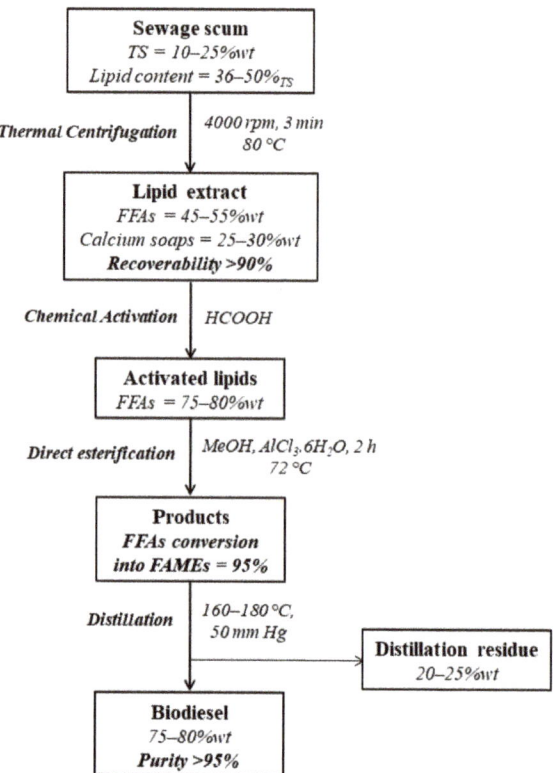

Figure 2. Scheme of biodiesel synthesis and purification from sewage scum.

2.2. Optimization of Transesterification Conditions for the Conversion of ME into Me-10-HSA and FAMEs

The conversion of ME into Me-10-HSA and FAMEs (according to the scheme reported in Figure 1) was optimized using AlCl$_3$·6H$_2$O and HCl as catalysts. According to the Box–Behnken experimental design described in Section 3.6 experiments were conducted on the distillation residue to find the optimal reaction conditions and study the process parameters' effect in the transesterification reaction. Experimental and predicted values for ME conversion at the design points are reported in Table 1.

A quadratic regression model was used to fit the experimental data, by obtaining the following relationships between factors and response for the two catalysts (Equations (1) and (2)):

$$ME\ conversion\ AlCl_3·6H_2O\ (\%) = -45.6167 + 11.6267C + 38.4455cat + 1.75413T +$$
$$+0.410478t - 1.01146C^2 - 34.8958cat^2 - 0.00761458T^2 - 0.00848126t^2 - 2.875Ccat +$$ (1)
$$-0.030625CT + 0.0644231Ct + 0.203125catT + 0.0865385catt - 0.00134615Tt$$

$$ME\ conversion\ HCl\ (\%) = 57.6733 + 10.5721C + 33.0761cat - 0.0389744T +$$
$$+0.684689t - 0.404167C^2 - 10.4948cat^2 + 0.00192708T^2 - 0.000690335t^2 - 2.4375Ccat +$$ (2)
$$-0.054375CT + 0.00384615Ct - 0.034375catT - 0.177885catt - 0.00442308Tt$$

The graphs between the predicted and the experimental ME conversion (%) reported in Figure 3 show that expected values are similar to the observed values, therefore validating the model's reliability in establishing the correlation between the process variables and the ME conversion.

Table 1. Box–Behnken design matrix for the four independent variables and the experimental ME conversion (%) using AlCl$_3$·6H$_2$O and HCl as catalysts.

E	Methanol (mL)	Catalyst (mmol)	Temperature (°C)	Time (h)	ME Conversion (%) AlCl$_3$·6H$_2$O		HCl	
					Pred.	Exp.	Pred.	Exp.
1	0	0	−1	−1	88.2	89.0	93.1	93.3
2	1	0	−1	0	94.2	94.5	86.2	84.3
3	0	1	−1	0	87.5	89.4	93.9	95.6
4	−1	0	−1	0	89.9	90.5	96.7	96.9
5	0	−1	−1	0	81.5	80.2	98.4	98.6
6	0	0	−1	1	85.3	84.3	89.3	88.6
7	0	1	0	−1	86.3	87.7	89.7	89.6
8	1	0	0	−1	95.2	94.7	95.4	95.3
9	0	−1	0	−1	94.5	92.8	95.2	96.3
10	−1	0	0	−1	87.2	87.7	96.7	96.9
11 [a]	0	0	0	0	92.9	93.4	95.3	95.5
12	−1	−1	0	0	87.7	86.6	99.9	98.1
13	1	1	0	0	97.1	97.9	94.8	93.3
14 [a]	0	0	0	0	89.9	90.5	99.6	99.7
15	1	−1	0	0	83.0	82.6	96.9	98.1
16 [a]	0	0	0	0	94.2	93.9	96.7	96.4
17	−1	1	0	0	89.3	89.2	98.7	98.5
18	0	−1	0	1	77.3	76.6	98.8	98.5
19	−1	0	0	1	80.7	82.7	98.3	98.1
20	0	1	0	1	97.2	97.5	94.8	93.3
21	1	0	0	1	86.7	85.3	97.0	96.2
22	0	0	1	−1	94.2	94.1	96.8	98.7
23	0	1	1	0	87.0	87.8	97.2	97.6
24	−1	0	1	0	94.3	95.4	99.3	99.0
25	0	−1	1	0	83.6	83.5	98.9	98.2
26	1	0	1	0	97.2	97.5	95.3	95.5
27	0	0	1	1	78.5	79.9	92.9	92.8

[a] Denoted as central points. Pred. = predicted values, Exp. = experimental values.

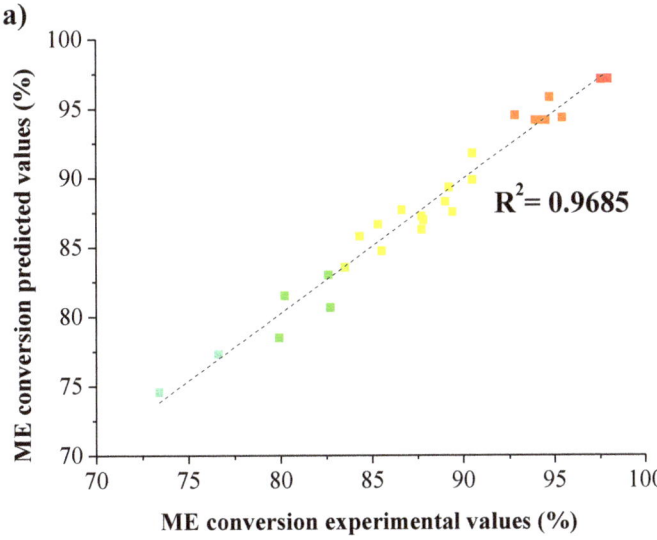

Figure 3. Cont.

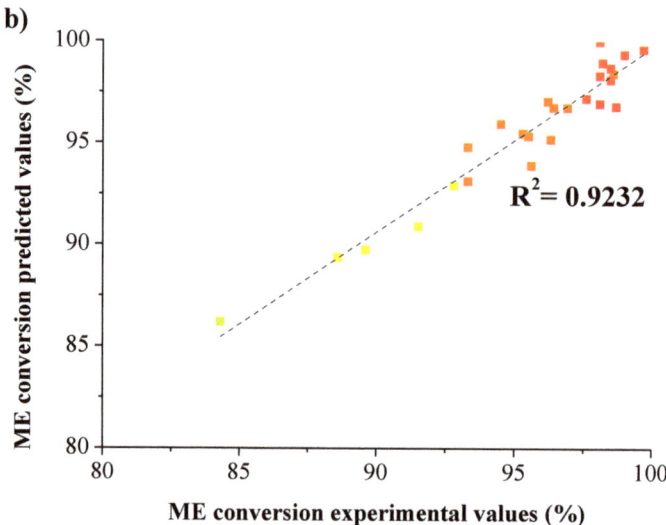

Figure 3. Predicted vs. experiment value for ME conversion (%) using (**a**) AlCl$_3$·6H$_2$O and (**b**) HCl as catalysts.

Subsequently, the significance of each parameter was evaluated by the analysis of variance (ANOVA) followed by Fisher's statistical test (F-test) for linear, interaction, and quadratic parameters in the second-order polynomial equations. In this work, the significance of the mathematical model adopted was associated with the *p*-value. A value of 0.05 was considered a suitable threshold with the corresponding significant parameters highlighted with an asterisk. The model's main statistics and the components of the fitting equations are given in Tables 2 and 3.

Table 2. The ANOVA summary table for the conversion of ME into Me-10-HSA and FAMEs using AlCl$_3$·6H$_2$O as catalyst.

Source	Sum of Squares	Df	Mean Square	F-Ratio	*p*-Value
Model	731.455	14	182.864	14.02	0.0000 *
C	167.253	1	167.253	64.72	0.0000 **
cat	181.741	1	181.741	70.33	0.0000 **
T	272.653	1	272.653	105.51	0.0000 **
t	109.808	1	109.808	42.49	0.0000 **
Ccat	21.16	1	21.16	8.19	0.0143 **
CT	6.0025	1	6.0025	2.32	0.1534
Ct	11.2225	1	11.2225	4.34	0.0592
catT	10.5625	1	10.5625	4.09	0.0661
catt	0.81	1	0.81	0.31	0.5859
Tt	0.49	1	0.49	0.19	0.6710
C^2	87.3001	1	87.3001	33.78	0.0001 **
cat^2	166.259	1	166.259	64.34	0.0000 **
T^2	49.4779	1	49.4779	19.15	0.0009 **
t^2	10.957	1	10.957	4.24	0.0619
Total error	31.0092	12	2.5841		
Total (corr.)	1018.31	26			

R^2 = 96.95% * $p < 0.05$ indicates model is significant. R^2(adjusted for d.f.) = 93.39% ** $p < 0.05$ indicates model terms are significant

Table 3. The ANOVA summary table for the conversion of ME into Me-10-HSA and FAMEs using HCl as catalyst.

Source	Sum of Squares	Df	Mean Square	F-Ratio	p-Value
Model	236.882	14	59.2204	11.79	0.0000 *
C	82.6875	1	82.6875	35.10	0.0001 **
cat	86.4033	1	86.4033	36.67	0.0001 **
T	36.75	1	36.75	15.60	0.0019 **
t	31.0408	1	31.0408	13.17	0.0035 **
Ccat	15.21	1	15.21	6.46	0.0259 **
CT	18.9225	1	18.9225	8.03	0.0151 **
Ct	0.04	1	0.04	0.02	0.8985
catT	0.3025	1	0.3025	0.13	0.7263
catt	3.4225	1	3.4225	1.45	0.2513
Tt	5.29	1	5.29	2.25	0.1599
C^2	13.9393	1	13.9393	5.92	0.0316 **
cat^2	15.0379	1	15.0379	6.38	0.0266 **
T^2	3.16898	1	3.16898	1.35	0.2687
t^2	0.0725926	1	0.0725926	0.03	0.8636
Total error	28.2725	12	2.35604		
Total (corr.)	347.365	26			

R^2 = 93.32% * $p < 0.05$ indicates model is significant. R^2(adjusted for d.f.) = 90.46% ** $p < 0.05$ indicates model terms are significant.

The p-associated values for the models adopted were less than 0.05, indicating that the model used to describe the transesterification reaction of ME with methanol was statistically significant. All linear parameters were substantial in the transesterification process. In particular, it was noted that there was a considerable difference in the relationship between the independent variables and their effects on the response variable (ME conversion) for the two catalysts. Using $AlCl_3 \cdot 6H_2O$ as a catalyst, temperature showed the most significant impact in the transesterification reaction followed by the amount of the catalyst, methanol and reaction time (Table 2). Instead, in the case of HCl, the amount of catalyst and methanol were the most significant variables with respect to temperature and reaction time (Table 3). As for the other terms (interaction and quadratic parameters), only Ccat (the interaction between the amount of methanol and catalyst), C^2 (the quadratic term associated with the methanol amount) and cat^2 (the quadratic term related to the amount of catalyst) were significant for both catalysts. Then, the goodness of fit of the models was checked by the coefficient of determination R^2. The value obtained of 0.9339 and 0.9046, respectively, for $AlCl_3 \cdot 6H_2O$ and HCl in its adjusted form, confirmed the efficacy of the model adopted.

Finally, response surface plots were generated to investigate the influence of the process parameters in the conversion of ME and identify the optimal experimental conditions required for both catalysts. Figure 4a shows the combined effect of methanol and catalyst, at a fixed temperature of 100 °C and a reaction time of 17 h. By increasing the amount of methanol and catalyst, an increase of conversion of ME was observed: 95 and 100%, were respectively obtained for $AlCl_3 \cdot 6H_2O$ and HCl with 5 mL of methanol and 1 mmol of catalyst. In Figure 4b, the effect of the temperature with the reaction time was investigated (methanol = 3 mL, catalyst = 0.6 mmol). In this case, the key role played by temperature in the transesterification reaction for the two systems studied is clear. At 120 °C, for $AlCl_3 \cdot 6H_2O$, the transesterification process's kinetic was particularly slow and long reaction times (30 h) were required for the full conversion of ME. In contrast, the reaction was complete after a few hours (4–6 h) using HCl. Figure 4c,d show the combined effect of temperature and catalyst (methanol = 3 mL, time = 17 h) and reaction time and methanol (catalyst = 0.6 mmol, temperature = 100 °C), respectively. These combinations of factors positively influenced the conversion of ME, obtaining a value close to 100%.

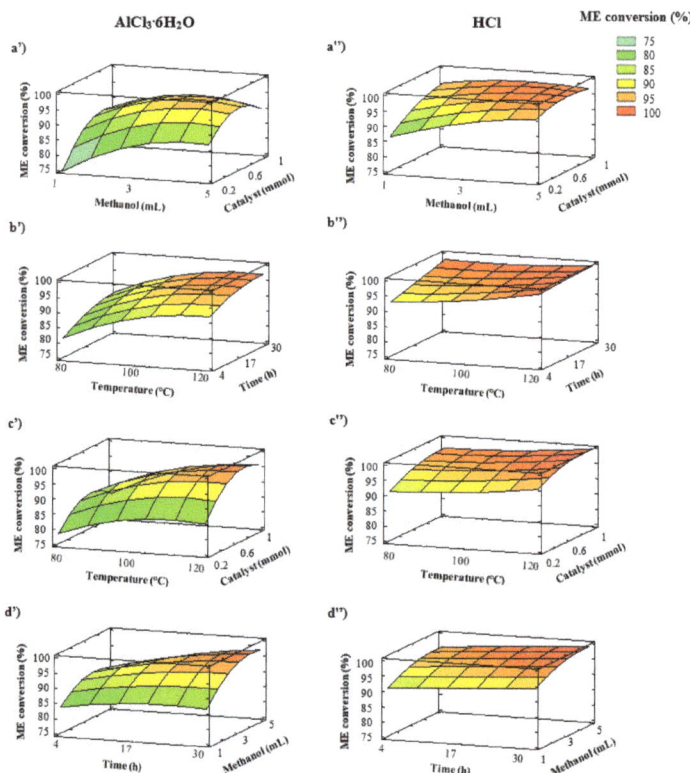

Figure 4. Response surface plot of the combined effects of: (**a**) methanol and catalyst amount (temperature = 100 °C, time = 17 h), (**b**) temperature and reaction time (methanol = 3 mL, catalyst = 0.6 mmol), (**c**) temperature and catalyst (methanol = 3 mL, time = 17 h), (**d**) time and methanol (catalyst = 0.6 mmol, temperature = 100 °C) for $AlCl_3 \cdot 6H_2O$ and HCl.

Based on these results, the optimal conditions were determined and directly applied in the transesterification of ME with methanol. The results obtained are reported in Table 4. Predicted responses are found to be in good agreement with the experimental results. In detail, using $AlCl_3 \cdot 6H_2O$ as a catalyst (0.76 mmol), a ME conversion of 99.6% was obtained at 115 °C after 30 h of reaction and 3.9 mL of methanol. Instead, the reaction catalyzed by HCl (1 mmol) was much faster: a ME conversion of 99.8% was achieved at 120 °C with a reduced amount of methanol (2.1 mL) after 4 h. In the absence of the catalyst, using the highest amount of methanol (5 mL) at 120 °C, a ME conversion of only 3% was obtained, confirming the efficiency of both catalysts.

Table 4. Results of the model validation under optimum conditions using $AlCl_3 \cdot 6H_2O$ and HCl as catalysts.

Catalysts	Methanol (mL)	Amount (mmol)	Temperature (°C)	Time (h)	ME Conversion (%)	
					Pred.	Exp.
$AlCl_3 \cdot 6H_2O$	3.9	0.76	115	30	99.4	99.6
HCl	2.1	1	120	4	100	99.8

Pred. = predicted values, Exp. = experimental values.

The transesterification in methanol of stearyl stearate was also evaluated to compare the different reactivity between HCl and $AlCl_3 \cdot 6H_2O$ vs. a fatty ester less congested sterically.

As can be seen from Table 5, for both the catalysts, a total conversion of stearyl stearate in methyl stearate was obtained at 100 °C after 24 h. In the case of HCl catalysis, the conversion is high even after 2 h at 70 °C (80%), showing a higher reaction rate than that of $AlCl_3 \cdot 6H_2O$, for which the conversion was only about 20%. Furthermore, regarding $AlCl_3 \cdot 6H_2O$, the results obtained clearly show that the amounts of methanol and catalyst greatly influence the transformation of stearyl stearate. This different behavior could be ascribed to the different strength of acidity among these two catalysts and the higher steric hindrance related to the hexa-aquo complex of aluminum chloride [45]. In fact, it was already demonstrated that the partial substitution of water coordinated to the aluminum center in $AlCl_3 \cdot 6H_2O$ produces a mixed-aquo-alcohol complex, which acts as a Brønsted acid.

Table 5. Stearyl stearate conversions under HCl and $AlCl_3 \cdot 6H_2O$ catalysis.

	HCl		$AlCl_3 \cdot 6H_2O$	
	70 °C, 2 h	100 °C, 24 h	70 °C, 2 h	100 °C, 24 h
Conversion (%)	80	100	20	100

2.3. Analysis of the Reaction Products

Once identified the optimal experimental conditions required for the complete conversion of ME, the organic phase was processed by column chromatography [28] and the products obtained were isolated, analyzed and quantified (Figures S1–S8). Based on 100 g of distillation residue, 28.2 g FAMEs and 32.3 g Me-10-HSA were respectively achieved. A detailed analysis of the fatty acids (FAs) profile was carried out by comparing the chemical composition of the methyl esters obtained with that of biodiesel previously recovered from FAME distillation. The results are reported in Figure 5.

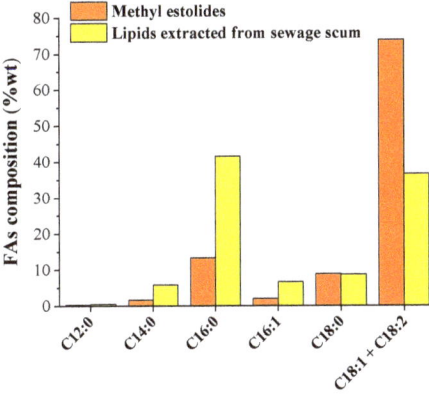

Figure 5. Comparison of FAs profile (%wt.) of methyl estolides and lipids extracted from sewage scum.

ME contained predominately oleic acid (C18:1) and linoleic acid (C18:2), totaling 73.9% of fatty acids present (AMW FAs = 276.8 g/mole). Instead, the distilled biodiesel from sewage scum (Figure 2) showed nearly equal amounts of saturated fatty acid and unsaturated fatty acid (AMW FAs = 266.8 g/mole) (Figure S9). This difference in FAs profile can be attributed to a possible origin of estolides as reaction products between 10-HSA (obtained from the enantioselective microbial hydration of oleic acid [25]) the subsequent esterification and/or transesterification with the oils and fats present in the stream. Considering only the market value of biodiesel produced (EUR 0.8 kg^{-1} [58]), a potential gain of EUR 225 could be obtained for each ton of the sample treated, leading to a further enhancement of sewage sludge to produce biofuels and biochemicals. However, the greatest profits would be obtained from Me-10-HSA produced during the process.

A high enantiomeric excess (ee > 92%) of R-enantiomeric form was observed (Figure S10), compound employed as the precursor to produce biochemicals, namely γ-(R)-dodecalactone. Since its current market value ranges from EUR 800 to 3000 kg^{-1} [59], the economy of the whole process would be greatly improved. As described above, the synthesis of 10-HSA generally requires the use of enzymes. The use of HCl or $AlCl_3 \cdot 6H_2O$ as a catalyst not only is significantly cheaper (EUR 0.8 kg^{-1} [60]), but in the case of $AlCl_3 \cdot 6H_2O$ at the end of its use, it could potentially be used in WWTPs as a coagulant, further contributing to the overall economy of the process.

3. Materials and Methods

3.1. Reagents and Instruments

All chemical reagents used in this work were of analytical grade and were used directly without further purification or treatment. Hexane (C_6H_{14}, 99%), toluene (C_6H_6, 99%), methanol (CH_3OH, 99.8%) and methyl heptadecanoate ($C_{18}H_{36}O_2$, ≥99%) were purchased from Sigma-Aldrich. Aluminum chloride hexahydrate ($AlCl_3 \cdot 6H_2O$, 99%) was obtained as pure-grade reagent from Baker. Ethanol (C_2H_5OH, ≥99.8%), diethyl ether (($C_2H_5)_2O$, 99%), formic acid (HCOOH, 99%), sulfuric acid (H_2SO_4, 98%), hydrochloric acid (HCl, 37%) and potassium hydroxide (KOH, 85%) were purchased from Carlo Erba.

A Rotofix 32 Hettich Centrifuge was used for the centrifugation experiments.

Identification of ME, Me-10-HSA and FAMEs was carried out by gas chromatography-mass spectroscopy (GC-MS) using a Perking Elmer Clarus 500 equipped with a Clarus spectrometer. Quantitative determinations were performed using a Varian 3800 GC-FID. Helium was used as a carrier gas with a flow of 1.3 mL min^{-1}. Both instruments were configured for cold on-column injections with a HP-5MS capillary column (30 m; Ø 0.32 mm; 0.25 µm film). The same temperature program was employed for the injector and the oven. The initial temperature was set to 60 °C and kept constant for 2 min. Then, it increased to 300 °C with a 15 °C min^{-1} ramp and kept constant for other 20 min. The temperature of detector (FID) was set to 300 °C. For GC-MS, the ion source was set to 70 eV and maintained at 250 °C.

FTIR spectra were recorded by a Perkin Elmer FTIR Spectrum BX instrument using KBr cells (neat compounds).

^{1}H NMR spectra were recorded on a Bruker AV-400 spectrometer using the residual solvent peak as a reference [61].

3.2. Sewage Scum

Sewage scum was collected from WWTPs of Bari West (240,000 Population Equivalent, PE), located in South of Italy. Samples were immediately processed to avoid long storage time (within two days, 4 °C) and characterized in terms of total solids, lipids, proteins, cellulose, lignin and ashes [29].

3.3. Experimental Procedure for Lipids Characterization

3.3.1. Determination of FFAs and Soaps

FFAs were determined by titration of the acidity present with a 0.1 N KOH solution and phenolphthalein (≥99%, Sigma-Aldrich) as an indicator. A total of 1 g of the sample collected was previously dissolved into 150 mL of a 1:1 v/v diethyl ether:ethanol mixture. Using the same experimental conditions, soaps were determined by titration with a 0.1 N HCl solution and methyl red (99%, Sigma-Aldrich) as an indicator.

3.3.2. Determination of Fatty Acids Profile and Average Molecular Weight

In a glass Pyrex reactor of 5 mL, 0.02 g of sample were dissolved with 2 mL of a 2:2:0.01 $v/v/v$ toluene:methanol:concentrated H_2SO_4 solution. The system was closed and placed into an ultrasonic bath at 70 °C for 5 h. Then, 1 mL of methyl heptadecanoate toluene solution (1000 ppm) was added as internal standard and the resulting solution gas-chromatographically analyzed (1 µL). Average molecular weight (AMW) was determined according to the following equation (Equation (3)):

$$AMW = \frac{\sum A_i MW_i}{\sum A_i} \quad (3)$$

where A_i and MW_i are the area and molecular weight of FFAs identified, respectively. Then, FAMEs content (%wt.) was calculated respect to methyl heptadecanoate as follows (Equation (4)):

$$\text{FAMEs content} = \frac{\sum A_i}{A_{sdt}} \times \frac{m_{sdt}}{m_{sample}} \times 100 \quad (4)$$

where A_{sdt} and m_{sdt} are the area and the mass of standard (methyl heptadecanoate), respectively, and m_{sample} is the amount of sample analyzed.

3.4. Extraction of Lipid Fraction from Sewage Scum and Chemical Activation

In a glass Pyrex reactor of 250 mL, 100 g of sewage scum were placed and closed. The system was heated in an oven at 80 °C. After this thermal treatment, the sample was rapidly centrifuged at 4000 rpm for 3 min by obtaining a three-phasic system consisting of: (i) an upper organic brown oily phase, (ii) a lower phase of wet residual solid, and (iii) an aqueous intermediate phase. The oily phase was recovered and stored at 4 °C for the subsequent operations. The isolated product was mainly constituted by FFAs (51.7%wt) and calcium soaps of fatty acids (30.4%wt). Then, the stoichiometric amount of HCOOH respect to the calcium soaps was added (4.8 g for 100 g of raw lipids) and the activated lipids recovered as clear oil, after centrifugation (4000 rpm, 1 min) at 80 °C [44].

3.5. Conversion of Activated Lipids into Methyl Esters of Recovery of Biodiesel Produced by Distillation Process

Activated lipids extracted from sewage scum were converted into the corresponding methyl esters, by direct esterification with methanol using $AlCl_3 \cdot 6H_2O$ as a catalyst [45]. The reaction was carried out at 72 °C for 2 h with a molar ratio FFAs:MeOH:catalyst = 1:10:0.02. At the end of the process, the reagent mixture was cooled to room temperature with the formation of bi-phasic system consisting of: (i) a light methanol layer (in which the catalyst was present) and (ii) a lower oily layer composed of methyl esters. The oily phase was recovered, and the residual methanol removed under vacuum (60 °C, 700 mmHg). Finally, biodiesel (75–80%wt, purity > 99%) was collected by subsequently vacuum distillation (160–180 °C, 50 mm Hg). The distillation residue (20–25%wt) was instead recovered, dried under nitrogen flow and analyzed by column chromatography [29], obtaining the chemical composition reported in Table 6.

Table 6. Chemical composition of the distillation residue obtained after direct esterification of the lipid fraction with methanol and recovery of biodiesel produced by distillation process.

Chemical Species	Composition (%wt.)
Mineral oil	2.7
Waxes	1.8
FAMEs	6.4
Me-10-HAS	3.6
Methyl-10-ketostearate	0.4
Methyl estolides	50.3
Acids	1.0
Other polar compounds	33.8

3.6. Transesterification Reaction of Methyl Estolides (or Stearyl Stearate) with Methanol

In a typical reaction, 0.1 g of sample (the distillation residue recovered in the previous step, ME content = 50.3% or stearyl stearate) were placed with methanol (1 mL) and 0.2 mmol of catalyst (HCl or $AlCl_3 \cdot 6H_2O$) in a glass Pyrex reactor of 15 mL. The system was closed and placed into a thermostatic bath at 80 °C for 4 h under agitation (250 rpm), using a magnetic stirring. Then, it was cooled to room temperature and the residual methanol was removed under nitrogen flow. Where it was possible, the catalyst was recovered by centrifugation and the organic phase was analyzed by gas chromatography for the determination of FAMEs content. Following the same procedure, Me-10-HSA and ME were also determined with the calibration curves obtained from the pure product, previously isolated by column chromatography [28].

Optimization of Transesterification Conditions

A three-step approach was used to investigate the effects of the process variables in the conversion of ME into Me-10-HSA and FAMEs and maximize their yield [44]. Methanol (C) and catalyst (cat) amount, temperature (T) and reaction time (t) were selected as independent variables (factors), while methyl estolides (ME) conversion was set as dependent variable (response). The experimental range of the levels and the independent variables considered in this study are presented in Table 7.

Table 7. Experimental range and levels of independent variables.

Variables	Symbol	Range and Levels			
		Lower Level (−1)	Center Level (0)	Upper Level (+1)	ΔX_i [a]
Methanol (mL)	C	1	3	5	2
Catalyst (mmol)	cat	0.2	0.6	1	0.4
Temperature (°C)	T	80	100	120	20
Time (h)	t	4	17	30	13

[a] Step change values.

A total of 27 experiments (including three replicates for the center point), were used for fitting a second-order response surface. The effects of factors on the response were analyzed according to the following quadratic function (Equation (5)):

$$Y = \beta_0 + \sum_{i=1}^{n} \beta_i X_i + \sum_{i=1}^{n} \beta_{ii} X_i^2 + \sum_{i=1}^{n}\sum_{i<j}^{n} \beta_{ij} X_i X_j \qquad (5)$$

where Y represents the ME conversion (%), X_i and X_j are the independent variables, β_0, β_i, β_{ij} and β_{ii} are the offset term, linear, interaction, and quadratic parameters, respectively. Statgraphycs® Centurion XVI was used for the regression analysis and the plot response surface. Then, to verify the validation of the overall fit of the developed regression model, the data obtained were processed by the analysis of variance (ANOVA). The adequacy of the polynomial model to fit experimental data were expressed as R^2 (coefficient of determination) and in its adjusted form. The statistical significance of R^2 was checked by the F-test at a confidence level of 95%.

Finally, the optimization of reaction conditions was carried out using response surface methodology (RSM) combined with the desirability function approach to form the desirability optimization methodology (DOM) [45].

4. Conclusions

In this work, for the first time, an efficient method was proposed for the synthesis of Me-10-HSA and FAMEs by direct conversion of methyl estolides isolated from sewage scum. A response surface methodology was applied to investigate the effect of the process variables on the methyl estolides conversion and maximize the final yield. $AlCl_3 \cdot 6H_2O$

and HCl were used as effective catalysts in promoting transesterification with methanol. HCl is more active in promoting the transesterification of methyl estolides to produce FAMEs and Me-(R)-10-HSA: a total conversion was in fact obtained already after 4 h. In the case of $AlCl_3 \cdot 6H_2O$, under similar reactive conditions, 20 h were necessary to achieve a conversion of 99.4%. On the other side, $AlCl_3 \cdot 6H_2O$ is a solid catalyst, easy to manage and less corrosive than mineral conventional acids [62]. $AlCl_3 \cdot 6H_2O$-catalyzed reaction resulted principally affected by temperature, whereas in the case of HCl, the amount of catalyst and methanol were the most significant variables. The obtainment of Me-(R)-10-HSA, in its (almost) pure enantiomeric form, increases the potential of sewage scum. For the specific case of $AlCl_3 \cdot 6H_2O$, the possible final use of the relevant residues in WWTPs as a coagulant results in a new scheme of valorization of a special waste, namely sewage scum, in which no secondary waste was generated. The transesterification of methyl estolides could actually implement the scenario of the full valorization of sewage scum towards a multi-products biorefinery. With the use of a limited number of reagents namely MeOH and $AlCl_3 \cdot 6H_2O$, and an integrated network of processes, biodiesel, methyl estolides, and Me-(R)-10-HSA would be effectively obtained from sewage scum in a sustainable way.

Supplementary Materials: The following are available online at https://www.mdpi.com/article/10.3390/catal11060663/s1, Figure S1: GC-chromatogram of methyl estolides isolated by distillation process, Figure S2: FTIR spectra of methyl estolides, Figure S3: GC-MS chromatogram of 10-(palmitoyloxy)-stearic methyl ester, Figure S4: GC-MS chromatogram of 10-(stearoyloxy)-stearic methyl ester, Figure S5: GC-chromatogram of methyl 10-(R)-hydroxystearate isolated, Figure S6: FTIR spectra of methyl 10-(R)-hydroxystearate, Figure S7: GC-MS chromatogram of methyl 10-(R)-hydroxystearate, Figure S8: ^{1}H NMR of isolated methyl 10-(R)-hydroxystearate, Figure S9: Comparison of chromatographic profiles of FAs obtained from methyl estolides and lipids extracted from sewage scum, Figure S10: Chemical structures of derivatizing agents (D1, D2 and D3) used for the determination of the absolute configuration of the Methyl 10-Hydroxy stearic acid (M10-HSA) isolated from sewage scum.

Author Contributions: L.d.B.: Methodology, Writing—original draft, Investigation, Validation. V.D.: Methodology, Writing—original draft, Investigation. C.P.: Conceptualization, Methodology, Investigation, Resources, Writing—review & editing, Supervision, Project administration, Funding acquisition. All authors have read and agreed to the published version of the manuscript.

Funding: This study was financially supported by MIUR (ERANETMEDNEXUS-14-035 Project WE-MET). This research was partially supported by European Union—FESR "PON Ricerca e Innovazione 2014–2020. Progetto: Energie per l'Ambiente TARANTO—Cod. ARS01_00637".

Data Availability Statement: The data presented in this study are contained within the article and the supplementary material.

Conflicts of Interest: The authors declare no conflict of interest.

Abbreviations

Roman Letters

A_i	Gas-chromatographic area of fatty acids detected
$AlCl_3 \cdot 6H_2O$	Aluminum Chloride Hexahydrate
A_{std}	Area of standard (methyl heptadecanoate)
C	Amount of Methanol
Cat	Amount of catalyst ($AlCl_3 \cdot 6H_2O$ or HCl)
C_2H_5OH	Ethanol
$C_{18}H_{36}O_2$	Methyl heptadecanoate
CH_3OH	Methanol
$(C_2H_5)_2O$	Diethyl ether
C_6H_{14}	Hexane
C_7H_8	Toluene

FAMEs	Fatty Acid Methyl Esters
FAs	Fatty Acids
FFAs	Free Fatty Acids
HCOOH	Formic acid
HCl	Hydrochloric Acid
HFAs	Hydroxy Fatty Acids
10-HAS	10-(R)-Hydroxystearic acid
H_2SO_4	Sulfuric acid
KOH	Potassium hydroxide
m_{sample}	Mass of sample analyzed
m_{std}	Mass of standard (methyl heptadecanoate)
ME	Methyl Estolides
Me-10-HAS	Methyl 10-(R)-Hydroxystearate
MW_i	Molecular weight of fatty acids detected
T	Temperature
TS	Total Solids
T	Time
X_i, X_j	Independent variables
Y	Dependent variable
Greek Letters	
β_0	Offset term
$\beta_i, \beta_{ij}, \beta_{ii}$	Linear, interaction, and quadratic parameters

References

1. Löwe, J.; Gröger, H. Fatty Acid Hydratases: Versatile Catalysts to Access Hydroxy Fatty Acids in Efficient Syntheses of Industrial Interest. *Catalysts* **2020**, *8*, 287. [CrossRef]
2. Cao, Y.; Zhang, X. Production of long-chain hydroxy fatty acids by microbial conversion. *Appl. Microbiol. Biotechnol.* **2013**, *97*, 3323–3331. [CrossRef] [PubMed]
3. Kolar, M.J.; Konduri, S.; Chang, T.; Wang, H.; McNerlin, C.; Ohlsson, L.; Härröd, M.; Siegel, D.; Saghatelian, A. Linoleic acid esters of hydroxy linoleic acids are anti-inflammatory lipids found in plants and mammals. *J. Biol. Chem.* **2019**, *294*, 10698–10707. [CrossRef] [PubMed]
4. Rodríguez, J.P.; Guijas, C.; Astudillo, A.M.; Rubio, J.M.; Balboa, M.A.; Balsinde, J. Sequestration of 9-Hydroxystearic Acid in FAHFA (Fatty Acid Esters of Hydroxy Fatty Acids) as a Protective Mechanism for Colon Carcinoma Cells to Avoid Apoptotic Cell Death. *Cancers* **2019**, *11*, 524. [CrossRef] [PubMed]
5. Suriyamongkol, P.; Weselake, R.; Narine, S.; Moloney, M.; Shah, S. Biotechnological approaches for the production of polyhydroxyalkanoates in microorganisms and plants-A review. *Biotechnol. Adv.* **2007**, *25*, 148–175. [CrossRef]
6. Verlinden, R.A.; Hill, D.J.; Kenward, M.A.; Williams, C.D.; Radecka, I. Bacterial synthesis of biodegradable polyhydroxyalkanoates. *J. Appl. Microbiol.* **2007**, *102*, 1437–1449. [CrossRef]
7. Chen, Y.; Biresaw, G.; Cermak, S.C.; Isbell, T.A.; Ngo, H.L.; Chen, L.; Durham, A.L. Fatty Acid Estolides: A Review. *J. Am. Oil Chem. Soc.* **2020**, *97*, 231–241. [CrossRef]
8. Cecilia, J.A.; Plata, D.B.; Maria, R.; Saboya, A.; Murilo, F.; De Luna, T.; Cavalcante, C.L.; Rodríguez-castellón, E. An Overview of the Biolubricant Production Process: Challenges and Future Perspectives. *Processes* **2020**, *8*, 257. [CrossRef]
9. De Haro, J.C.; Garrido, M.D.P.; Ángel, P.; Carmona, M.; Rodríguez, J.F. Full conversion of oleic acid to estolides esters, biodiesel and choline carboxylates in three easy steps. *J. Clean. Prod.* **2018**, *184*, 579–585. [CrossRef]
10. McNutt, J.; He, Q. (Sophia) Development of biolubricants from vegetable oils via chemical modification. *J. Ind. Eng. Chem.* **2016**, *36*, 1–12. [CrossRef]
11. Boratyński, F.; Szczepańska, E.; De Simeis, D.; Serra, S.; Brenna, E. Bacterial Biotransformation of Oleic Acid: New Findings on the Formation of γ-Dodecalactone and 10-Ketostearic Acid in the Culture of Micrococcus Luteus. *Molecules* **2020**, *25*, 3024. [CrossRef] [PubMed]
12. Kourist, R.; Hilterhaus, R. Microbial lactone synthesis based on renewable resources. In *Microorganisms in Biorefineries*; Springer: Berlin/Heidelberg, Germany, 2015; pp. 275–301.
13. An, J.-U.; Joo, Y.-C.; Oh, D.-K. New Biotransformation Process for Production of the Fragrant Compound γ-Dodecalactone from 10-Hydroxystearate by Permeabilized Waltomyces lipofer Cells. *Appl. Environ. Microbiol.* **2013**, *79*, 2636–2641. [CrossRef] [PubMed]
14. Soni, S.; Agarwal, M. Lubricants from renewable energy sources—A review. *Green Chem. Lett. Rev.* **2014**, *7*, 359–382. [CrossRef]
15. Yao, L.; Hammond, E.G.; Wang, T.; Bhuyan, S.; Sundararajan, S. Synthesis and physical properties of potential biolubricants based on ricinoleic acid. *J. Am. Oil Chem. Soc.* **2010**, *87*, 937–945. [CrossRef]
16. Hiseni, A.; Arends, I.W.C.E.; Otten, L.G. New Cofactor-Independent Hydration Biocatalysts: Structural, Biochemical, and Biocatalytic Characteristics of Carotenoid and Oleate Hydratases. *ChemCatChem* **2014**, *7*, 29–37. [CrossRef]

17. Jenkins, T.C.; AbuGhazaleh, A.A.; Freeman, S.; Thies, E.J. The Production of 10-Hydroxystearic and 10-Ketostearic Acids Is an Alternative Route of Oleic Acid Transformation by the Ruminal Microbiota in Cattle. *J. Nutr.* **2006**, *136*, 926–931. [CrossRef]
18. Chen, Y.Y.; Liang, N.Y.; Curtis, J.M.; Gänzle, M.G. Characterization of Linoleate 10-Hydratase of Lactobacillus plantarum and Novel Antifungal Metabolites. *Front. Microbiol.* **2016**, *7*, 1561. [CrossRef]
19. Kim, B.-N.; Joo, Y.-C.; Kim, Y.-S.; Kim, K.-R.; Oh, D.-K. Erratum to: Production of 10-hydroxystearic acid from oleic acid and olive oil hydrolyzate by an oleate hydratase from Lysinibacillus fusiformis. *Appl. Microbiol. Biotechnol.* **2012**, *95*, 1095–1096. [CrossRef]
20. Kim, K.-R.; Kang, W.-R.; Oh, D.-K. Complete genome sequence of Stenotrophomonas sp. KACC 91585, an efficient bacterium for unsaturated fatty acid hydration. *J. Biotechnol.* **2017**, *241*, 108–111. [CrossRef] [PubMed]
21. Hudson, J.A.; MacKenzie, C.A.M.; Joblin, K.N. Conversion of oleic acid to 10-hydroxystearic acid by two species of ruminal bacteria. *Appl. Microbiol. Biotechnol.* **1995**, *44*, 1–6. [CrossRef]
22. Kang, W.-R.; Seo, M.-J.; Shin, K.-C.; Park, J.-B.; Oh, D.-K. Gene cloning of an efficiency oleate hydratase fromStenotrophomonas nitritireducensfor polyunsaturated fatty acids and its application in the conversion of plant oils to 10-hydroxy fatty acids. *Biotechnol. Bioeng.* **2017**, *114*, 74–82. [CrossRef] [PubMed]
23. Kang, W.R; Seo, M.J.; Shin, K.C.; Park, J.B.; Oh, D.K. Comparison of biochemical properties of the original and newly identified oleate hydratases from Stenotrophomonas maltophilia. *Appl. Environ. Microbiol.* **2017**, *83*. [CrossRef]
24. Kuo, T.M.; Lanser, A.C.; Nakamura, L.K.; Hou, C.T. Production of 10-ketostearic acid and 10-hydroxystearic acid by strains of Sphingobacterium thalpophilum isolated from composted manure. *Curr. Microbiol.* **2000**, *40*, 105–109. [CrossRef]
25. Serra, S.; De Simeis, D.; Castagna, A.; Valentino, M. The Fatty-Acid Hydratase Activity of the Most Common Probiotic Microorganisms. *Catalysts* **2020**, *10*, 154. [CrossRef]
26. Hou, C.T. Biotechnology for fats and oils: New oxygenated fatty acids. *New Biotechnol.* **2009**, *26*, 2–10. [CrossRef] [PubMed]
27. Pera, L.M.; Baigori, M.D.; Pandey, A.; Castro, G.R. Biocatalysis. In *Industrial Biorefineries & White Biotechnology*; Elsevier: Amsterdam, The Netherlands, 2015; pp. 391–408. [CrossRef]
28. Di Bitonto, L.; Todisco, S.; Gallo, V.; Pastore, C. Urban sewage scum and primary sludge as profitable sources of biodiesel and biolubricants of new generation. *Bioresour. Technol. Rep.* **2020**, *9*, 100382. [CrossRef]
29. Di Bitonto, L.; Lopez, A.; Mascolo, G.; Mininni, G.; Pastore, C. Efficient solvent-less separation of lipids from municipal wet sewage scum and their sustainable conversion into biodiesel. *Renew. Energy* **2016**, *90*, 55–61. [CrossRef]
30. Di Bitonto, L.; Locaputo, V.; D'Ambrosio, V.; Pastore, C. Direct Lewis-Brønsted acid ethanolysis of sewage sludge for production of liquid fuels. *Appl. Energy* **2020**, *259*, 114163. [CrossRef]
31. Casiello, M.; Catucci, L.; Fracassi, F.; Fusco, C.; Laurenza, A.G.; Di Bitonto, L.; Pastore, C.; D'Accolti, L.; Nacci, A. ZnO/Ionic Liquid Catalyzed Biodiesel Production from Renewable and Waste Lipids as Feedstocks. *Catalysts* **2019**, *9*, 71. [CrossRef]
32. *Urban Waste Grease Resource Assessment*; NREL: Golden, CO, USA, 1998; NREL/SR-570-26141.
33. Igoni, A.H.; Ayotamuno, M.; Eze, C.; Ogaji, S.; Probert, S. Designs of anaerobic digesters for producing biogas from municipal solid-waste. *Appl. Energy* **2008**, *85*, 430–438. [CrossRef]
34. Bertanza, G.; Canato, M.; Laera, G.; Tomei, M.C. Methodology for technical and economic assessment of advanced routes for sludge processing and disposal. *Environ. Sci. Pollut. Res.* **2014**, *22*, 7190–7202. [CrossRef] [PubMed]
35. Chipasa, K.B.; Mdrzycka, K. Characterization of the fate of lipids in activated sludge. *J. Environ. Sci.* **2008**, *20*, 536–542. [CrossRef]
36. Di Bitonto, L.; Pastore, C. Up-grading of Waste Oil: A Key Step in the Future of Biofuel Production. In *Process Systems Engineering for Biofuels Development*; John Wiley & Sons: New York, NY, USA, 2020; pp. 121–147. [CrossRef]
37. Huang, Y.; Li, Y.; Han, X.; Zhang, J.; Luo, K.; Yang, S.; Wang, J. Investigation on fuel properties and engine performance of the extraction phase liquid of bio-oil/biodiesel blends. *Renew. Energy* **2020**, *147*, 1990–2002. [CrossRef]
38. Lawan, I.; Zhou, W.; Idris, A.L.; Jiang, Y.; Zhang, M.; Wang, L.; Yuan, Z. Synthesis, properties and effects of a multi-functional biodiesel fuel additive. *Fuel Process. Technol.* **2020**, *198*, 106228. [CrossRef]
39. Tamilselvan, P.; Nallusamy, N.; Rajkumar, S. A comprehensive review on performance, combustion and emission characteristics of biodiesel fuelled diesel engines. *Renew. Sustain. Energy Rev.* **2017**, *79*, 1134–1159. [CrossRef]
40. Singh, D.; Sharma, D.; Soni, S.; Sharma, S.; Sharma, P.K.; Jhalani, A. A review on feedstocks, production processes, and yield for different generations of biodiesel. *Fuel* **2020**, *262*, 116553. [CrossRef]
41. Tubino, M.; Junior, J.G.R.; Bauerfeldt, G.F. Biodiesel synthesis: A study of the triglyceride methanolysis reaction with alkaline catalysts. *Catal. Commun.* **2016**, *75*, 6–12. [CrossRef]
42. Kwon, E.E.; Kim, S.; Jeon, Y.J.; Yi, H. Biodiesel Production from Sewage Sludge: New Paradigm for Mining Energy from Municipal Hazardous Material. *Environ. Sci. Technol.* **2012**, *46*, 10222–10228. [CrossRef]
43. Mondala, A.; Liang, K.; Toghiani, H.; Hernandez, R.; French, T. Biodiesel production by in situ transesterification of municipal primary and secondary sludges. *Bioresour. Technol.* **2009**, *100*, 1203–1210. [CrossRef]
44. Pastore, C.; Lopez, A.; Mascolo, G. Efficient conversion of brown grease produced by municipal wastewater treatment plant into biofuel using aluminium chloride hexahydrate under very mild conditions. *Bioresour. Technol.* **2014**, *155*, 91–97. [CrossRef]
45. Pastore, C.; Barca, E.; Del Moro, G.; Lopez, A.; Mininni, G.; Mascolo, G. Recoverable and reusable aluminium solvated species used as a homogeneous catalyst for biodiesel production from brown grease. *Appl. Catal. A Gen.* **2015**, *501*, 48–55. [CrossRef]
46. Matsumura, S. Enzyme-Catalyzed Synthesis and Chemical Recycling of Polyesters. *Macromol. Biosci.* **2002**, *2*, 105–126. [CrossRef]
47. Ajith, B.; Math, M.; Gc, M.P.; Parappagoudar, M.B. Analysis and optimisation of transesterification parameters for high-yield Garcinia Gummi-Gutta biodiesel using RSM and TLBO. *Aust. J. Mech. Eng.* **2020**, 1–16. [CrossRef]

48. Ferella, F.; Di Celso, G.M.; De Michelis, I.; Stanisci, V.; Vegliò, F. Optimization of the transesterification reaction in biodiesel production. *Fuel* **2010**, *89*, 36–42. [CrossRef]
49. Kansedo, J.; Lee, K.T.; Bhatia, S. Biodiesel production from palm oil via heterogeneous transesterification. *Biomass Bioenergy* **2009**, *33*, 271–276. [CrossRef]
50. Silva, G.F.; Camargo, F.L.; Ferreira, A.L. Application of response surface methodology for optimization of biodiesel production by transesterification of soybean oil with ethanol. *Fuel Process. Technol.* **2011**, *92*, 407–413. [CrossRef]
51. Demirbas, A. Biofuels sources, biofuel policy, biofuel economy and global biofuel projections. *Energy Convers. Manag.* **2008**, *49*, 2106–2116. [CrossRef]
52. Escobar, J.C.; Lora, E.S.; Venturini, O.J.; Yáñez, E.E.; Castillo, E.F.; Almazan, O. Biofuels: Environment, technology and food security. *Renew. Sustain. Energy Rev.* **2009**, *13*, 1275–1287. [CrossRef]
53. Wang, Y.; Feng, S.; Bai, X.; Zhao, J.; Xia, S. Scum sludge as a potential feedstock for biodiesel production from wastewater treatment plants. *Waste Manag.* **2016**, *47*, 91–97. [CrossRef]
54. Kargbo, D.M. Biodiesel Production from Municipal Sewage Sludges. *Energy Fuels* **2010**, *24*, 2791–2794. [CrossRef]
55. Bi, C.-H.; Min, M.; Nie, Y.; Xie, Q.-L.; Lu, Q.; Deng, X.-Y.; Anderson, E.; Li, N.; Chen, P.; Ruan, R. Process development for scum to biodiesel conversion. *Bioresour. Technol.* **2015**, *185*, 185–193. [CrossRef] [PubMed]
56. Capodaglio, A.G.; Callegari, A. Feedstock and process influence on biodiesel produced from waste sewage sludge. *J. Environ. Manag.* **2018**, *216*, 176–182. [CrossRef] [PubMed]
57. Olkiewicz, M.; Torres, C.M.; Jiménez, L.; Font, J.; Bengoa, C. Scale-up and economic analysis of biodiesel production from municipal primary sewage sludge. *Bioresour. Technol.* **2016**, *214*, 122–131. [CrossRef]
58. Silva, J.F.L.; Grekin, R.; Mariano, A.P.; Filho, R.M. Making Levulinic Acid and Ethyl Levulinate Economically Viable: A Worldwide Technoeconomic and Environmental Assessment of Possible Routes. *Energy Technol.* **2018**, *6*, 613–639. [CrossRef]
59. Serra, S.; De Simeis, D. New insights on the baker's yeast-mediated hydration of oleic acid: The bacterial contaminants of yeast are responsible for the stereoselective formation of (R)-10-hydroxystearic acid. *J. Appl. Microbiol.* **2018**, *124*, 719–729. [CrossRef] [PubMed]
60. Schwiderski, M.; Kruse, A. Process design and economics of an aluminium chloride catalysed organosolv process. *Biomass Convers. Biorefinery* **2016**, *6*, 335–345. [CrossRef]
61. Fulmer, G.R.; Miller, A.J.M.; Sherden, N.H.; Gottlieb, H.E.; Nudelman, A.; Stoltz, B.M.; Bercaw, J.E.; Goldberg, K.I. NMR Chemical Shifts of Trace Impurities: Common Laboratory Solvents, Organics, and Gases in Deuterated Solvents Relevant to the Organometallic Chemist. *Organometallics* **2010**, *29*, 2176–2179. [CrossRef]
62. Pastore, C.; D'Ambrosio, V. Intensification of Processes for the Production of Ethyl Levulinate Using $AlCl_3 \cdot 6H_2O$. *Energies* **2021**, *14*, 1273. [CrossRef]

Article

MgO Catalysts for FAME Synthesis Prepared Using PEG Surfactant during Precipitation and Calcination

Valdis Kampars *, Ruta Kampare and Aija Krumina

Institute of Applied Chemistry, Faculty of Material Sciences and Applied Chemistry, Riga Technical University, LV-1048 Riga, Latvia; ruta.kampare@rtu.lv (R.K.); aija.krumina@rtu.lv (A.K.)
* Correspondence: valdis.kampars@rtu.lv

Abstract: To develop a method for the preparation of MgO nanoparticles, precatalyst synthesis from magnesium nitrate with ammonia and calcination was performed in presence of PEG in air. Without PEG, the catalysts are inactive. The conversion to hydroxide was performed using a PEG/MgO molar ratio of 1, but, before the calcination, excess of PEG was either saved (PEG1) or increased to 2, 3, or 4 (PEG 2–4). Catalysts were calcined at 400–660 °C and characterized using XRD, N_2 adsorption-desorption, TGA, FTIR, and SEM. The FAME yield in the reactions with methanol depend on the PEG ratio used and the calcination temperature. The optimal calcination temperature and highest FAME yield in the 6 h reactions for catalysts PEG1, PEG2, PEG3 and PEG4 were 400 °C, 74%; 500 °C, 80%; 500 °C, 51% and 550 °C, 31%, respectively. The yield dependence on calcination temperature for catalysts with a constant PEG ratio is similar to that of a bell curve, which becomes wider and flatters with an increase in PEG ratio. For most catalysts, the FAME yield increases as the size of the crystallites decreases. The dependence of FAME and the intermediate yield on oil conversion confirms that all catalysts have strong base sites.

Keywords: heterogeneous catalysts; combustion; PEG; transesterification; biodiesel

1. Introduction

Biodiesel is one of the three main biofuels and is the main biofuel in the EU. The European biodiesel market has been stabilized at around 10 million tons/year [1]. Biodiesel is currently produced from high-quality vegetable oil (EU rapeseed oil) by transesterification with methanol at 65 °C in the presence of a homogeneous basic catalyst using a methanol/triglyceride molar ratio of 6/9. Reaction time does not exceed 1h. In parallel with the production of biofuels from high-quality edible oils, there is a growing trend towards the use of cheap and non-edible raw materials for the production of advanced biodiesel [2]. The RED II Directive provides for a gradual transition for Member States from conventional to advanced biofuels, with the following shares of energy consumption in the transport sector: 0.2% in 2022, 1% in 2025, and 3.5% in 2030 [3]. The production of advanced biofuels makes it possible to use local raw materials, including waste, and to reduce the extremely harmful effects of the transportation sector on the environment and climate.

Regardless of the raw material, the currently dominant acquisition process involves esterification in the presence of a homogeneous catalyst, and has various significant drawbacks:

- The process can be carried out if the triglyceride does not contain more than 1–2% free fatty acids (FFA);
- The separation of biodiesel from glycerol and glycerides is not fully realized;
- The quality of both biodiesel and the by-product glycerol is low and the treatment of crude products polluting the environment is necessary;
- The catalyst material is not reusable and enters the product and the purification system;

- The production of biodiesel from low-quality feedstock requires pre-treatment and the resulting raw material has increased the content of free fatty acids (FFA), so the use of homogeneous catalyst causes soap formation.

All the above-mentioned problems could be solved by using heterogeneous basic catalysts [4,5]. The development of these catalysts was intensified by the RED II Directive requirements, but the achieved results so far are unsatisfactory. Transesterification reactions in the presence of active basic heterogeneous catalysts usually take place at 90–150 °C for 4–8 h [6,7]. The only heterogeneous catalyst developed by the AXENS corporation, which has proven to be good and durable in industrial operations, works at a temperature of 190–220 °C and pressure of 40–70 bar [8]. As the currently dominant biodiesel production technology is implemented at atmospheric pressure, it is not expected that developments will take place in terms of the complexity of the production infrastructure and there is a significant increase in production cost by using high temperature and pressure. Undoubtedly, the most promising direction of development in the industry is processing at low temperature in the presence of more active and robust heterogeneous catalysts.

Biodiesel synthesis from triglyceride (TG) proceeds accordingly to the balanced summary equation (Equation (1)):

$$TG + 3\ MeOH \rightleftharpoons 3\ FAME + G \qquad (1)$$

where MeOH—methanol; FAME—mixture of fatty acid methyl esters (biodiesel); and G—glycerol.

Reaction (1) consists of three reversible stages and proceeds with the production of intermediates, such as diglycerides (DG) and monoglycerides (MG). In the presence of homogeneous basic catalysts, reaction is initiated by generation of methoxide anion (2), which attacks the carbonyl group of TG and causes fast FAME production without losing the active anion intermediates:

$$MeOH + B^- \rightleftharpoons MeO^- + BH \qquad (2)$$

where B^- are HO^- or RO^-.

All reactions are reversible, therefore, the partitioning of glycerol in the form of separate layers is very important for shifting the equilibrium (1) to the right. In order for the reactions to proceed rapidly and selectively, a necessary minimum B^- concentration must be maintained [9]. In the presence of basic heterogeneous catalysts, the mechanism is considered to be similar, but the role of the base is to create strong base sites on the surface of the catalysts. Similar to homogeneous catalysis, the reaction is initiated by methoxide anion generation and the activity of the heterogeneous catalyst should be determined by the density of the strong base sites on its surface [10–13]. Most of the successfully investigated metal oxide basic catalysts for transesterification reactions belong to the M(II)O group and are ionic. The surface of these catalysts is terminated by metal cations and oxide anions (O^{2-}), and contains different types of defects and environments (kinks, steps, terraces), which play a determining role in the catalytic phenomenon [14]. Successfully tested catalysts for biodiesel production are BeO, MgO, CaO, SrO, and BaO [15]. From these, calcium oxide is favored as it is highly active, cheap, and eco-friendly [16]. Unfortunately, this catalyst is leached during reaction and loses its activity in contact with air, it therefore cannot be considered as a prospective choice [15]. More stable and promising is MgO, which is used in medicine as a catalyst and an adsorbent [17,18]. The polarizing power of magnesium ion (3.9) is remarkably stronger than that of calcium ion (2.0), which would make the building of strong base sites more difficult than in case of CaO [19]. This is confirmed by the high dependence of MgO catalyst activity on manufacturing conditions [20]. Solution combustion synthesis of catalysts increases activity by employing metal salts as oxidants and different organic compounds as fuels in order to use the heat of the redox reaction [21–23]. The characteristic stages of catalyst synthesis according to this method are: sol solution, viscous gel, dried gel, and calcined product preparation.

The most active oxidants are nitrates, but their mixture with organic materials is explosive and there are less prospective industrial uses. To prevent the explosiveness proceeding the calcination stage, conversion from the nitrates to the hydroxides is widely used [17]. The present work is undertaken to investigate the synthesis of active MgO catalysts, and avoiding explosion during calcination by replacing strong oxidants with a weak oxidant, such as air.

2. Results and Discussion
2.1. Preparation of Catalysts and Their Characteristics

Magnesium salts, such as nitrate, sulfate, chloride, carbonate, acetate, and citrate, are most often used as feedstocks for the synthesis of MgO. Often, these salts are not used directly as a source of magnesium, but are converted into a magnesium hydroxide precatalyst. This is based on the considerations that low-temperature decomposition of magnesium hydroxide yields pure magnesium oxide. Thus, it is possible to perform calcination without harmful gases. The addition of different organic compounds as fuels before calcination prevents the formation of explosive mixtures. Thermal decomposition of magnesium hydroxide is well investigated. Its low decomposition temperature (300–400 °C) allows easy activation of the precatalyst [20,24]. As presented in Figure 1, the calcination of $Mg(OH)_2$ at 375 °C is sufficient for obtaining the polycrystalline cubic structure of MgO nanoparticles.

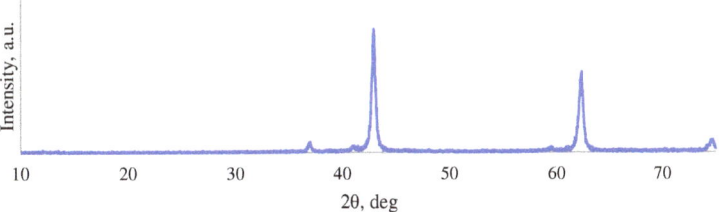

Figure 1. XRD patterns of PEG1, calcined at 375 °C for 3 h, obtained using Cu K-alpha radiation (0.15406 nm).

The peaks at 37°, 43°, 62.4°, and 74.7° correspond to (111), (200), (220), and (311) reflections, which reveal the formation of the polycrystalline structure of MgO nanoparticles [25]. No peaks from impurities were detected in the registered patterns. The sharp diffraction peaks indicate the good crystallinity of the MgO products.

It is known that higher activity of catalysts is obtained when fuel and surfactants are present in both precatalyst synthesis and in the activation stages [26–28]. During the preparation of a precatalyst, PEG 6000 works as a surfactant, but during the calcination it is as a surfactant and fuel. Our research shows that the presence of PEG during the synthesis of the precatalyst in a molar ratio of MgO of 1 creates a small effect, but a further increase in PEG shows no improvement up to 4. Therefore, in all cases, we performed the first step using a molar ratio of PEG 6000 to MgO of 1. The presence of surfactant and fuel at the calcination stage changes the structure and activity of the resulting catalyst and the effect depends on the ratio to catalyst [28]. Thermal analysis up to 700 °C (Figure 2) of precatalysts was performed to examine PEG 6000 effects in an inert atmosphere.

From the TG curves, it can be seen that samples PEG1 and PEG4 display a very notable mass loss near 400 °C, which is related to the decomposition of PEG 6000. In the row PEG1–PEG2–PEG3–PEG4, gradual changes in the TG curves occur, suggesting that the calcination temperature at 400 °C would be sufficient to obtain active MgO nanoparticles.

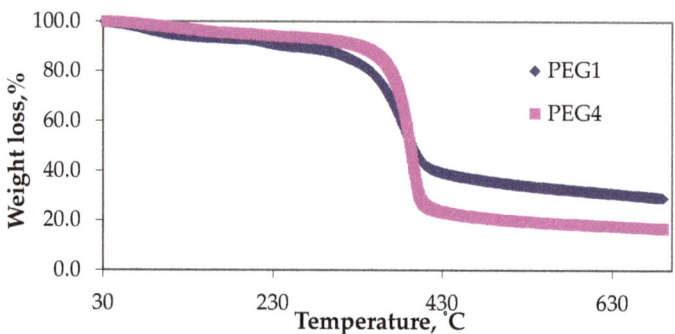

Figure 2. Thermogravimetric analysis of precursors of PEG1 and PEG4.

As shown in Figure 2, a slight weight loss is still observed in the temperature range of 430 °C to 730 °C, which can be explained by the elimination of adsorbed carbonate species [20].

The nitrogen adsorption–desorption isotherms and specific surface area, total pore volume, the pore sizes of the catalysts are shown in Figure 3 and Table 1, respectively. All catalysts have large pores with an average pore diameter of around 5 nm. As seen from Figure 3, the isotherms of MgO are type IV with an H3 hysteresis loop.

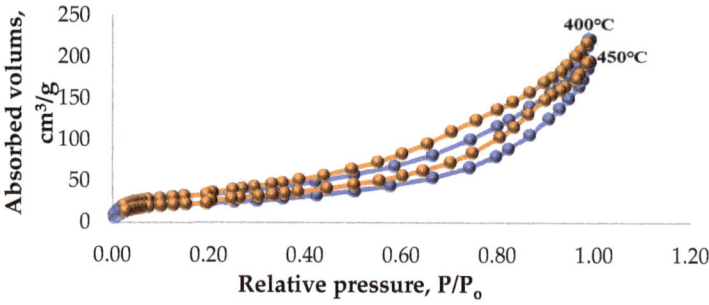

Figure 3. Nitrogen adsorption–desorption isotherms of PEG2 calcined at 400 °C and 450 °C.

Table 1. Textural properties of MgO.

Catalyst	Total Surface Area, m²/g	Pore Volume, mL/g	Pore Size, nm	FAME Yield, %
PEG2 400	126.2	0.366	4.95	42.7
PEG2 450	85.7	0.319	4.98	73.9

As seen in Table 1, the change in the textural characteristics is small and the number of performed experiments does not allow to draw conclusions about the effect of texture on the activity of the catalysts.

Characteristic SEM images of the obtained MgO are presented in Figure 4. Two different forms of aggregates dominate. These images indicate that the powder sample is an agglomeration of nanoparticles. Irregular aggregated flake-like particles with a rough surface represented the active catalysts, while MgO composed of many irregular nanoparticles did not catalyze the interesterification reactions. A similar phenomenon, with a significant morphological effect on the activity of catalysts or adsorbents, has also been observed in other works [29,30]. The aggregation process with the formation of a unique morphology may play a crucial role in ensuring catalytic activity.

 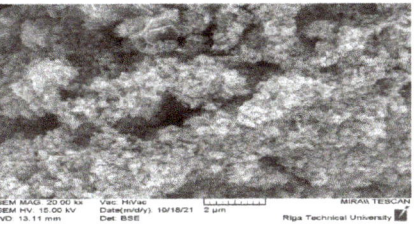

Figure 4. SEM image of active (**left**) and inactive (**right**) MgO catalysts.

2.2. Catalytic Performance

Initialization of the transesterification reaction proceeded according to Equation (2). The catalytic activity of heterogeneous catalysts depends on the density of strong base sites on the catalyst surface. This depends on the morphology, number, and types of defects [10–13]. The presence of PEG improved MgO activity. Catalysts synthesized without PEG under experimental conditions did not provide a FAME yield above 3%.

2.2.1. Effects of PEG Content and Calcination Temperature

To determine the effect of PEG content and calcination temperature on catalytic performance, each PEG1–PEG4 precatalyst was calcined at a different temperature. As shown in Figure 5, sharp maxima of the FAME yield were registered for the PEG1 catalyst at 400 °C, for the PEG2 catalyst at 500 °C, for the PEG3 catalyst at 500 °C, and there was a flatter maximum for PEG4 at 550 °C.

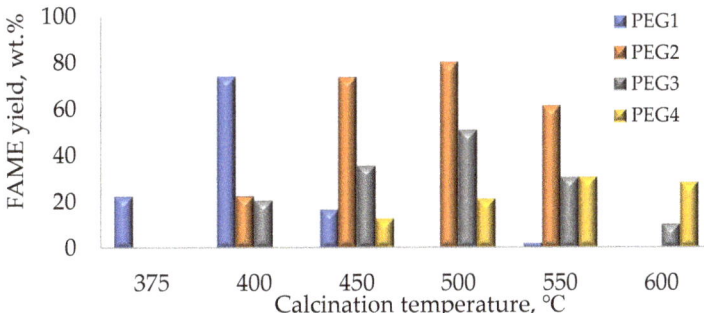

Figure 5. Effect of PEG content and calcination temperature on FAME yield in transesterification reactions at 65 °C after 6h.

The repetition of individual points showed that the standard deviations of the FAME yield of the PEG1 catalyst at 400 °C reached 15.5%. It also showed that the density of strong base catalytic sites depends on more than just the calcination temperature. It was also affected by minor changes in the calcination process, including air circulation, crucible form, the volume of precatalyst in the crucible, and location in the muffle furnace. As the combustion process proceeds by the reaction of fuel with oxygen in the air, this seems understandable and requires a more precise process for providing airflow control. The standard deviation of the results was lowered by increasing the PEG content, but a PEG content above 3 lowered the activity. The shape of the distribution of the experimental points from the calcination temperature for the catalyst with a constant PEG ratio is similar to a bell curve and becomes wider and lower with an increase in PEG content. PEG2 provides the best catalysts.

The dependence of the activity of MgO catalysts on small changes in the manufacturing process is well known and the safe production of an active MgO catalyst for practical use is indeed a serious problem. Babak et al. concluded that unsupported MgO was not a suitable catalyst, but Li et al. showed that mesoporous-supported MgO catalyst only becomes active at 220 °C [31,32]. At the same time, works on very high activity MgO catalysts have been published [6,26,27]. The problem of MgO catalyst activity has been analyzed using specific surface properties in the form of research and quantum mechanics calculation methods. Di Cosimo et al. confirmed that the surface of the magnesium oxide catalyst obtained by the calcination of magnesium hydroxide contains three types of sites with different basicity: surface sites of strong (low coordination O^{2-} anions), medium (oxygen in Mg^{2+}-O^{2-} pairs), and weak (OH- groups) basicity [20]. Decomposition of $Mg(OH)_2$ at 400 °C generates hydroxylated MgO, mainly containing strong O^{2-} basic sites located in surface defects, such as the corners and edges of the crystalline solid surface. An increase in the calcination temperature removes the OH groups and also surface solid defects, creating more stable structures that contains a higher concentration of medium-strength Mg^{2+}-O^{2-} basic pair sites. Thus, an increase in the calcination temperature drastically decreased the density of strong base sites and, to a lesser extent, that of weak OH groups, while it slightly increased that of medium-strength base sites; therefore, a calcination temperature of 400 °C seems to be optimal [20]. According to Montero et al., water and methanol chemisorb preferentially over defects and edge sites over NanoMgO-700 through the conversion of surface O^{2-} sites to OH^- and the coincident creation of Mg-OH or Mg-OCH_3 moieties, respectively [33]. According to this work, the best calcination temperature for magnesium hydroxide would be 700 °C [33]. It could be considered that both publications indicate the area fir the best calcination temperatures and the results for PEG catalysts, which narrow this area to 400–600 °C (Figure 5).

According to Vedrine, the main role in creating of strong base sites is different for different types of defects [14]. If this is so, then reducing the size of the crystallites should increase the density of the defects and also catalyst activity. As follows from Figure 6, in general, the FAME yield and, thus, the density of strong base sites, increase as the size of the crystallites decreases. This could indicate an association between the defects of the boundary surfaces of these crystallites with the number of strong basic centers.

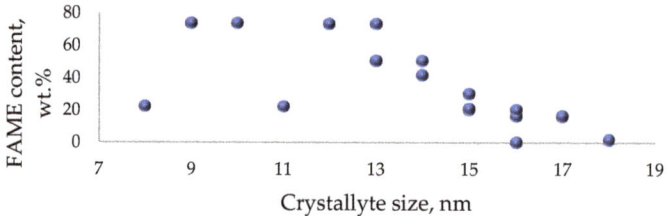

Figure 6. Relationship between FAME yield and crystallite size. The main crystallite size is calculated using (200) and (220) reflections using the Scherrer equation (Table S1).

However, two catalysts fall out of this relationship (PEG1, FAME content 22.5 wt.% and PEG2, FAME content 73.9 wt.%), which confirms that the size of the crystallites does not always determine the density of strong base sites, and other phenomena can rule out these relationships. An unexpected decrease in catalyst activity could be caused by partial blocking of the MgO surface with combustion products, or products resulting from incomplete calcination. To determine if this was the case, FTIR spectra were recorded for individual catalysts. The obtained FTIR spectra confirmed that not all catalysts have a clean MgO surface (Figure 7), although XRD and SEM-EDS analyses did not indicate this. Absorption bands near 1420 are assignable to magnesium carbonate or hydromagnesite [34], which is not uncommon for oxide catalysts. Therefore, catalyst surfaces control, determined with FTIR, should be included in determining the characteristics [35].

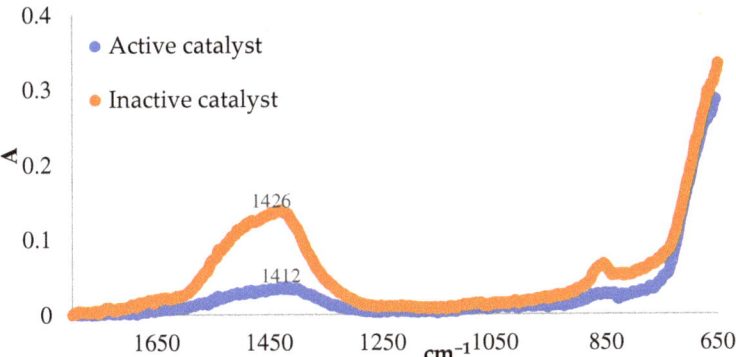

Figure 7. FTIR spectra of active (PEG2 crystallite size 13 nm, FAME 73.5%) and inactive (PEG1 crystallite size 8 nm, FAME 22.5%) catalysts.

2.2.2. Strong Base Sites and Reaction Proceeding

In reality, reaction (1) proceeds in three reversible stages:

$$TG + MeOH \rightleftharpoons FAME + DG \qquad (3)$$

where DG are diglycerides,

$$DG + MeOH \rightleftharpoons FAME + MG \qquad (4)$$

where MG are monoglycerides, and

$$MG + MeOH \rightleftharpoons FAME + G \qquad (5)$$

where G is glycerol.

Chromatographic analysis shows that the main constituents of the obtained reaction mixtures were FAME, MG, DG, and TG, if the amount of G did not exceed 1 wt.%. The sum of these components was usually 97–100%, which allowed us to consider that the amount of unknown by-product was insignificant. Assuming that the reaction mechanism was provided by the same active sites in all cases, the change in the composition of the reaction products depending on oil conversion should reflect the gradual realization of the transesterification. As follows from Figure 8, the experimental points form logical curves that confirm the presence of the same active sites in all cases.

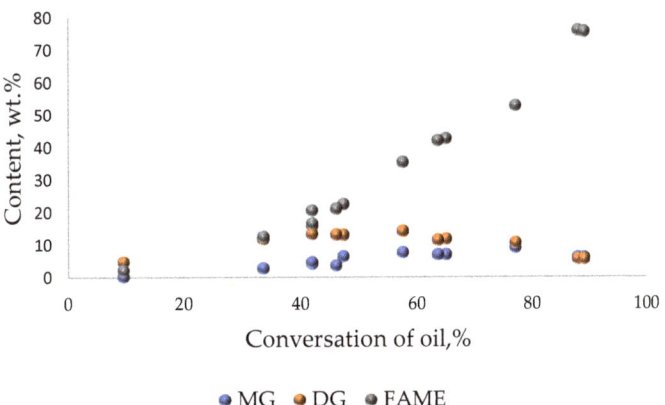

Figure 8. Composition of reaction mixture as a function on the conversion of oil.

The contents of MG and DG go through a maximum and approach zero above 95% conversion. DG content reaches a maximum with oil conversion of about 55%. However, the maximum content of MG was observed at an oil conversion of 70%. Up to 40% oil conversion, FAME content increases relatively slowly. As oil conversion increases, FAME content grows faster. In the conversion region, 40–80% FAME content increases at least twice as efficient than below 40%. Most likely, this is due to changes in the reaction rate of the reverse reactions caused by the gradual release of the glycerol in a separate layer.

3. Materials and Methods

3.1. Materials

Refined rapeseed oil was purchased from a local producer Iecavnieks & Co Ltd (Iecava, Latvia). The average molecular weight of the oil was 896 g/mol, density–0.92 g/mL at 20 °C, saponification value–186.7 mg KOH/g, and acid value–0.12 mg KOH/g. The percentages of fatty acids in the oil were: palmitic–4.1%, stearic–1.4%, oleic–62.5%, linoleic–21.7%, linolenic–8.7%, arachidic–0.4%, and other fatty acids–1.2 wt.%. The content of MG (monoglycerides) was 0.0 wt.%, DG (diglycerides)–0.4 wt.%, and TG–99.6 wt.%.

Magnesium nitrate hexahydrate, ammonia solution 25%, PEG 6000 and methanol were supplied by Sigma-Aldrich (St. Louis, MO, USA). Materials for GC analysis—1,2,4-butanetriol (96%) and MSTFA (N-Methyl-N-(trimethylsilyl)trifluoroacetamide (97%)–were purchased from Alfa Aesar (Haverhill, MA, USA), tricaprin (98%)—from TCI Europe (2070 Zwijndrecht, Belgium), pyridine (99.5%)—from Lach-Ner (Neratovice, Czech Republic), and dichloromethane (99.5%) was supplied by Chempur (Karlsruhe, Germany). Grade 1Whatman filter paper was supplied by Sigma-Aldrich.

3.2. Catalyst and Biodiesel Synthesis

Catalyst synthesis: 100 mL of 1M magnesium nitrate solution (25.6 g/100 mL) was added to a 250 mL Erlenmeyer flask and 4.4 g of PEG 6000 was then dissolved. The obtained mixture was stirred at 500 rpm for 1 h (homogenization stage N1). Then, it was filtrated with a NH_4OH (25% NH_3 basis) solution in a burette to adjust the pH up to 10. The NH_4OH solution was dripped from the burette slowly in a drop-wise manner. After reaching pH 10, the second homogenization process continued for 2h by stirring (homogenization stage N2) and then hydrothermal treatment was performed by leaving the solution overnight. The material obtained after centrifugation was suspended in 100 mL of PEG 6000 solution (PEG1 4.4 g/100 mL; PEG2 8.8 g/100 mL; PEG3 13.2 g/100 mL and PEG4 17.6 g/100 mL) using an ultrasound environment. Water was slowly removed using a rotary evaporator. The obtained material was placed in oven at 80 °C for 5 h to remove excess water moisture. The dried sample was placed in a crucible, which was placed inside a muffle furnace, for calcination at the selected temperature for 3 h. The temperature of the furnace was increased at the very slow rate of 1 °C·min^{-1} until the desired temperature of calcination was attained. Fine amorphous MgO nanoparticles were prepared and collected.

Biodiesel synthesis: 0.64 g of MgO catalyst (7% by weight of oil) was added to a 100 mL round bottom flask containing 11 mL of methanol (methanol to oil molar ratio of 27). The mixture was allowed to reflux at 65 °C for 0.5 h at 600 rpm, using a hotplate with a magnetic stirrer. Then, 10 mL of oil (10.3 mmol) diluted with 10 mL of co-solvent THF was added and refluxed at 65 °C at 600 rpm for 6 h. The obtained product was allowed to settle for 3 h. The crude biodiesel was separated from glycerol using a separating funnel. FAME was filtered using Whatman filter paper and, after removing the THF and excess methanol, stored in a glass container for analysis.

3.3. Catalyst Characterization

The samples were out-gassed at 150 °C for 24 h before measurements. The total surface area was estimated using the Brunauer–Emmett–Teller (BET) method. Pore diameters were derived from desorption isotherms using the Barrett–Joyner–Halenda (BJH) method.

XRD analysis was performed using an X-ray diffractometer (Bruker AXS D8 ADVANCE, Billerica, MA, USA) in the range of 10–75° (2θ) with Cu K-alpha radiation (0.15406 nm) and a step size of 0.02°.

The surfaces of the catalysts were studied with a Tescan MIRA3 LMU (Tescan, Brno, Czech Republic) scanning electron microscope. Prior to analysis, samples were coated with a gold layer using an Emitech K550X sputter coater.

Universal attenuated total reflectance-Fourier transform infrared spectroscopy (UATR-FTIR) was used for surface control of the catalysts. Measurements were carried out on a *PerkinElmer Spectrum 100* spectrometer connected to a *Universal ATR Sampling Accessory*. A spectral range 650–4000 cm^{-1} was selected.

TGA analysis was performed using a thermogravimetric analyzer (PerkinElmer STA 6000). The heating rate was 10 °C/min and experiments were performed under a pure nitrogen flow of 20 mL/min, supervised using a mass-flow controller.

3.4. Biodiesel Characterization

Analysis of all components from each sample of raw biodiesel was carried out by using an Analytical Controls (Rotterdam, The Netherlands) biodiesel analyser, based on the Agilent Technologies (Santa Clara, CA, USA) gas chromatograph 7890A. All compounds were analyzed using DB5-HT column (15 m, 0.32 mm, 0.10 µm) under conditions similar to those prescribed by standard EN 14105. Glycerol (G), monoglicerides (MG), diglicerides (DG), triglycerides (TG), and FAME were quantified as in our previous report [36]. The oven temperature was initially set to 50 °C for 5 min. Then, the temperature was first increased to 180 °C at a rate of 15 °C/min, then, to 230 °C at a rate of 7 °C/min, and, finally, to 370 °C at a rate of 10 °C/min. Helium was used as a carrier gas and the detector temperature was set to 390 °C. The reaction mixture was characterized by the mass percent content of the mentioned groups of compounds. The standard deviation for all analyzed compounds was not higher than 0.5 wt.%, but FAME and TG were not higher than 0.9 wt.%.

All experiments were repeated triplicate in order to determine the variability of the results and to assess experimental errors in GC analysis. Arithmetic averages and standard deviations were calculated for all results. Statistical analyses were performed using Microsoft Excel (2013). The yield of FAME was assumed to be equal to the wt.% of FAME content in the reaction mixture after removing glycerol, methanol, and the co-solvent.

4. Conclusions

From the alkaline earth metal oxides, MgO can be considered as the most promising catalyst for biodiesel synthesis. The chemical bond in MgO is less ionic than that in CaO, which makes it more stable but poses problems in the formation of a high density of strong base sites on the surface of nanoparticles to ensure a high catalytic activity. It is possible to significantly increase the activity of MgO nanoparticles using PEG as a surfactant and fuel, as well as with an optimal calcination temperature. FAME yield dependence on calcination temperature for catalysts obtained in presence of a constant molar PEG/MgO ratio is like a bell curve, the width, height, and position of which depend on that molar ratio. Increasing the PEG/MgO ratio has an optimum catalyst activity at 2. With the presence of catalysts PEG1, PEG2, PEG3, and PEG4, the maximum yield of FAME was achieved as 400 °C, 74%; 500 °C, 80%; 500 °C, 51%; and 550 °C, 31%, respectively. The narrowest zone of high activity was for the PEG1 catalyst and the widest and flattest was for PEG4. The PEG2 catalyst remaining the most prospective.

For most of the catalysts, the FAME yield increased as the size of the crystallites of catalyst decreased. The use of FTIR spectra showed that deviations in this relationship may be due to the retention of incomplete calcination products on the surface of MgO. FAME and intermediate yield dependence on oil conversion confirmed that all catalysts had the same types of strong base sites that were necessary for initialization of transesterification reactions.

Supplementary Materials: The following supporting information can be downloaded at: https://www.mdpi.com/article/10.3390/catal12020226/s1, Table S1. Reaction products by using different catalysts; Figure S1. XRD analysis of PEG1 catalyst Calcined at 400 °C; Figure S2. XRD analysis of PEG2 catalyst Calcined at 400 °C.

Author Contributions: Conceptualization, funding and final approval V.K.; methodology V.K., R.K. and A.K.; original draft preparation, editing V.K. and R.K., performed experiments and data collection R.K. and A.K. All authors have read and agreed to the published version of the manuscript.

Funding: This work was conducted as the project of the Latvian Council of Science lzp-2020/2-0194.

Data Availability Statement: Data are contained within the article and the Supplementary Materials.

Conflicts of Interest: The authors declare no conflict of interest.

References

1. 2022 European Biomass Industry Association. Available online: http://www.eubia.org/cms/wiki-biomass/biofuels/biodiesel/ (accessed on 10 February 2022).
2. Toldrá-Reig, F.; Mora, L.; Toldrá, F. Trends in Biodiesel Production from Animal Fat Waste. *Appl. Sci.* **2020**, *10*, 3644. [CrossRef]
3. Directive (EU) 2018/2001 of the European Parliament and of the council of 11 December 2018, on the pro motion of the use of energy from renewable sources. *Off. J. Eur. Union* **2018**, *L328/82*.
4. Fattah, I.M.R.; Ong, H.C.; Mahlia, T.M.I.; Mofijur, M.; Silitonga, A.S.; Rahman, S.M.A.; Ahmad, A. State of the Art of Catalysts for Biodiesel Production. *Front. Energy Res.* **2020**, *8*, 101. [CrossRef]
5. Tran-Nguyen, P.L.; Ong, L.K.; Go, A.W.; Ju, Y.; Angkawijaya, A.E. Non-catalytic and heterogeneous acid/base-catalyzed biodiesel production: Recent and future developments. *Asia-Pac. J. Chem. Eng.* **2020**, *15*, e2490. [CrossRef]
6. Dawood, S.; Ahmad, M.; Ullah, K.; Zafar, M.; Khan, K. Synthesis and characterization of methyl esters from non-edible plant species yellow oleander oil, using magnesium oxide (MgO) nano-catalyst. *Mater. Res. Bull.* **2018**, *101*, 371–379. [CrossRef]
7. Rabie, A.M.; Shaban, M.; Abukhadra, M.R.; Hosny, R.; Ahmed, S.A.; Negm, N.A. Diatomite supported by CaO/MgO nanocomposite as heterogeneous catalyst for biodiesel production from waste cooking oil. *J. Mol. Liq.* **2019**, *279*, 224–231. [CrossRef]
8. Dimian, A.C.; Bildea, C.S.; Kiss, A.A. *Technology & Engineering Applications in Design and Simulation of Sustainable Chemical Processes*; Elsevier: Amsterdam, The Netherlands, 2019; pp. 719–734. [CrossRef]
9. Kampars, V.; Abelniece, Z.; Lazdovica, K.; Kampare, R. Interesterification of rapeseed oil with methyl acetate in the presence of potassium tert-butoxide solution in tetrahydrofuran. *Renew. Energy* **2020**, *158*, 668–674. [CrossRef]
10. Sharma, S.; Saxena, V.; Baranwal, A.; Chandra, P.; Pandey, L.M. Engineered nanoporous materials mediated heterogeneous catalysts and their implications in biodiesel production. *Mater. Sci. Energy Technol.* **2018**, *1*, 11–21. [CrossRef]
11. Di Serio, M.; Tesser, R.; Pengmei, L.; Santacesaria, E. Heterogeneous Catalysts for Biodiesel Production. *Energy Fuels* **2008**, *22*, 207–217. [CrossRef]
12. Chouhan, A.S.; Sarma, A. Modern heterogeneous catalysts for biodiesel production: A comprehensive review. *Renew. Sustain. Energy Rev.* **2011**, *15*, 4378–4399. [CrossRef]
13. Refaat, A.A. Biodiesel production using solid metal oxide catalysts. *Int. J. Environ. Sci. Technol.* **2010**, *8*, 203–221. [CrossRef]
14. Védrine, J.C. Heterogeneous Catalysis on Metal Oxides. *Catalysts* **2017**, *7*, 341. [CrossRef]
15. Thangaraj, B.; Solomon, P.R.; Muniyandi, B.; Ranganathan, S.; Lin, L. Catalysis in biodiesel production—A review. *Clean Energy* **2019**, *3*, 2–23. [CrossRef]
16. Kesic, Ž.; Lukic, I.; Zdujic, M.; Mojovic, L.; Skala, D. Calcium oxide based catalysts for biodiesel production: A review. *Chem. Ind. Chem. Eng. Q.* **2016**, *22*, 391–408. [CrossRef]
17. Lin, J.; Nguyen, N.T.; Zhang, C.; Ha, A.; Liu, H.H. Antimicrobial Properties of MgO Nanostructures on Magnesium Substrates. *ACS Omega* **2020**, *5*, 24613–24627. [CrossRef]
18. Fernandes, M.; Singh, K.R.; Sarkar, T.; Singh, P.; Singh, R.P. Recent Applications of Magnesium Oxide (MgO) Nanoparticles in various domains. *Adv. Mater. Lett.* **2020**, *11*, 1–10. [CrossRef]
19. Available online: https://studylib.net/doc/7514936/polarizing-power-of-common-cations (accessed on 10 February 2022).
20. Di Cosimo, J.I.; Díez, V.K.; Ferretti, C.; Apesteguía, C.R. Basic catalysis on MgO: Generation, characterization and catalytic properties of active sites. *Catalysis* **2014**, *26*, 1–28. [CrossRef]
21. Balamurugan, S.; Ashna, L.; Parthiban, P. Synthesis of Nanocrystalline MgO Particles by Combustion Followed by Annealing Method Using Hexamine as a Fuel. *J. Nanotechnol.* **2014**, *2014*, 1–6. [CrossRef]
22. Li, F.-T.; Ran, J.; Jaroniec, M.; Qiao, S.Z. Solution combustion synthesis of metal oxide nanomaterials for energy storage and conversion. *Nanoscale* **2015**, *7*, 17590–17610. [CrossRef] [PubMed]
23. Li, S. Combustion synthesis of porous MgO and its adsorption properties. *Int. J. Ind. Chem.* **2019**, *10*, 89–96. [CrossRef]
24. Iwasaki, S.; Kodani, S.; Koga, N. Physico-Geometrical Kinetic Modeling of the Thermal Decomposition of Magnesium Hydroxide. *J. Phys. Chem. C* **2020**, *124*, 2458–2471. [CrossRef]
25. Nemade, K.R.; Waghuley, S.A. Synthesis of MgO Nanoparticles by Solvent Mixed Spray Pyrolysis Technique for Optical Investigation. *Int. J. Met.* **2014**, *2014*, 1–4. [CrossRef]

26. Abdulkadir, B.A.; Ramli, A.; Wei, L.J.; Uemura, Y. Effect of MgO Loading on the Production of Biodiesel from Jatropha Oil in the Presence of MgO/MCM-22 Catalyst. *J. Jpn. Inst. Energy* **2018**, *97*, 191–199. [CrossRef]
27. Margellou, A.; Koutsouki, A.; Petrakis, D.; Vaimakis, T.; Manos, G.; Kontominas, M.; Pomonis, P.J. Enhanced production of biodiesel over MgO catalysts synthesized in the presence of Poly-Vinyl-Alcohol (PVA). *Ind. Crop. Prod.* **2018**, *114*, 146–153. [CrossRef]
28. Cai, B.; Liu, H.; Han, W. Solution Combustion Synthesis of Fe_2O_3-Based Catalyst for Ammonia Synthesis. *Catalysts* **2020**, *10*, 1027. [CrossRef]
29. Vahid, B.R.; Haghighi, M.; Toghiani, J.; Alaei, S. Hybrid-coprecipitation vs. combustion synthesis of Mg-Al spinel based nanocatalyst for efficient biodiesel production. *Energy Convers. Manag.* **2018**, *160*, 220–229. [CrossRef]
30. Zhou, J.; Wang, W.; Cheng, Y.; Zhang, Z. Facile Hydrothermal Synthesis and Characterization of Porous Magnesium Oxide for Parachlorophenol Adsorption From the Water. *Integr. Ferroelectr.* **2012**, *137*, 18–29. [CrossRef]
31. Babak, S.; Iman, H.; Abdullah, A.Z. Alkaline Earth Metal Oxide Catalysts for Biodiesel Production from Palm Oil: Elucidation of Process Behaviors and Modeling Using Response Surface Methodology. *Iran. J. Chem. Chem. Eng.* **2013**, *32*, 113–126. [CrossRef]
32. Li, E.; Rudolph, V. Transesterification of Vegetable Oil to Biodiesel over MgO-Functionalized Mesoporous Catalysts. *Energy Fuels* **2007**, *22*, 145–149. [CrossRef]
33. Montero, J.M.; Isaacs, M.A.; Lee, A.F.; Lynam, J.M.; Wilson, K. The surface chemistry of nanocrystalline MgO catalysts for FAME production: An in situ XPS study of H_2O, CH_3OH and CH_3OAc adsorption. *Surf. Sci.* **2015**, *646*, 170–178. [CrossRef]
34. Li, S.; Wang, Z.J.; Chang, T.-T. Temperature Oscillation Modulated Self-Assembly of Periodic Concentric Layered Magnesium Carbonate Microparticles. *PLoS ONE* **2014**, *9*, e88648. [CrossRef]
35. Imani, M.M.; Safaei, M. Optimized Synthesis of Magnesium Oxide Nanoparticles as Bactericidal Agents. *J. Nanotechnol.* **2019**, *2019*, 1–6. [CrossRef]
36. Sustere, Z.; Murnieks, R.; Kampars, V. Chemical interesterification of rapeseed oil with methyl, ethyl, propyl and isopropyl acetates and fuel properties of obtained mixtures. *Fuel Process. Technol.* **2016**, *149*, 320–325. [CrossRef]

MDPI
St. Alban-Anlage 66
4052 Basel
Switzerland
Tel. +41 61 683 77 34
Fax +41 61 302 89 18
www.mdpi.com

Catalysts Editorial Office
E-mail: catalysts@mdpi.com
www.mdpi.com/journal/catalysts

www.ingramcontent.com/pod-product-compliance
Lightning Source LLC
LaVergne TN
LVHW070617100526
838202LV00012B/662